Praise for

THE GIANT LEAP

"*The Giant Leap* is a detailed and provocative exploration of what it means for life as we know it to escape the bounds of the only planet where it has ever been. We have taken the first tentative steps, and this book will inspire people to think more seriously about what the next ones can and should be."

—Sean Carroll, author of *The Biggest Ideas in the Universe*

"Caleb Scharf eloquently explains not only how difficult spaceflight is but also why it is so important to the future of humanity. Expanding into the Solar System will not be easy, but he shows how it will be essential, and perhaps inevitable."

—Jeff Foust, author at *SpaceNews*

THE
GIANT
LEAP

THE GIANT LEAP

WHY SPACE IS THE NEXT FRONTIER IN THE EVOLUTION OF LIFE

CALEB SCHARF

BASIC BOOKS
New York

Basic Books
Hachette Book Group
1290 Avenue of the Americas, New York, NY 10104
www.basicbooks.com

Printed in the United States of America

First Edition: October 2025

Published by Basic Books, an imprint of Hachette Book Group, Inc.
The Basic Books name and logo is a registered trademark of the Hachette Book Group.

The Hachette Speakers Bureau provides a wide range of authors for
speaking events. To find out more, go to hachettespeakersbureau.com
or email HachetteSpeakers@hbgusa.com.

Basic books may be purchased in bulk for business, educational, or promotional use.
For more information, please contact your local bookseller or the Hachette Book Group
Special Markets Department at special.markets@hbgusa.com.

The publisher is not responsible for websites (or their content) that are
not owned by the publisher.

Print book interior design by Bart Dawson.

Library of Congress Cataloging-in-Publication Data

Names: Scharf, Caleb, 1968– author.
Title: The giant leap : why space is the next frontier in the evolution of life / Caleb Scharf.
Description: First edition. | New York : Basic Books, 2025. |
Includes bibliographical references and index.
Identifiers: LCCN 2025019521 (print) | LCCN 2025019522 (ebook) |
ISBN 9781541604179 (hardcover) | ISBN 9781541604186 (ebook)
Subjects: LCSH: Interplanetary voyages. | Astronautics. | Evolution (Biology) |
Outer space—Exploration.
Classification: LCC QB500.262 .S34 2025 (print) | LCC QB500.262 (ebook) |
DDC 919.9/204—dc23/eng/20250510
LC record available at https://lccn.loc.gov/2025019521
LC ebook record available at https://lccn.loc.gov/2025019522

ISBNs: 9781541604179 (hardcover), 9781541604186 (ebook)

LSC-C

Printing 1, 2025

To Marina and Aaron Scharf, whose hearts were always with the stars

Earth is the cradle of humanity,

But one cannot live in a cradle forever.

—Konstantin Tsiolkovsky, "The Exploration of
Cosmic Space by Means of Reaction Devices,"
Vestnik Vozdukhoplavaniya 3 (1912)

CONTENTS

THE
GIANT
LEAP

—— 1 ——

OCEANS AND VESSELS

Both in space and time, we seem to be brought somewhat
near to that great fact—the mystery of mysteries—the first
appearance of new beings on this earth.

—Charles Darwin,
The Voyage of the Beagle (1839)[1]

We are a dying species. All species on Earth are. In a billion
years, our ever-brightening Sun will cause Earth's climate to
become a moist greenhouse that pumps water molecules high into
the stratosphere. There they will break apart into oxygen and hydrogen, and the hydrogen will escape to space, preventing those water
molecules from ever forming again and irrevocably drying the planet
out. A few billion years later, the Sun will inflate into a red giant star
whose outer reaches will drag the cinder that remains of the Earth
deeper into its stellar maw, ending and erasing all history of what once
took place here.[2] Even before these ultimate acts, the Earth will be
smashed again and again by massive asteroids with enough violent
force to purge most large and complex life from its continents and

oceans, as has already happened more than once across the past four billion years.[3]

Perhaps we're instinctively aware of these fraught conditions. Maybe that's why we can't help but look for some kind of cosmic salvation. And maybe this is why the velvety darkness of a clear night, experienced far from the pollution of human-made illumination, is a thing of such awe and terror: it reminds us of our entrapment but also of other possibilities. It's true that our eyes are sensitive enough to see only about two thousand stars in Earth's night skies, whereas the rest of the hundred billion that constitute our Milky Way galaxy are either too dim or too far, so they dissolve into a subdued haze. Yet those two thousand stellar points, scattered across the black canopy of Earth's sky, can induce a powerful feeling of vertigo, as if we might fall headlong into that yawning abyss. Our minds fill in the cavern of space with the pull of gravity and mystery even as we feel the comforting weight of our own bodies holding us to the ground.

Modern astronomical breakthroughs have told us that each of these other stars is also surrounded by its own set of planetary shores in an abundance of other worlds. These are the kinds of places that an exploring, or desperate, species might one day find itself washing up onto. All that has to happen to allow that beaching is for the void between to be traversed, and this journey, like most journeys, will always begin with one word: *launch*.

It is a simple word with great ambition, a word that has been used to describe everything from the gentlest of boating excursions to the most violent and explosive eruptions of missiles and rockets into the blue firmament and beyond. Across the centuries, people and objects have been launched into water, air, and space in aid of an endless parade of human endeavors and schemes. Even intangible things like ideas have been launched into existence, with many intended to co-opt or coerce others into buying wares or to craftily get them to embrace a philosophy or ideology. Other ideas are powerful enough on their own that, regardless of intent, their launch can wind up changing the collective human psyche.

But launches existed before the word existed and before humans existed. Life itself is an eruptive launch in the world, a pulsing fountain of self-directing matter. It swarms and grows, and it fills the niches that surround it and the niches that it creates seemingly out of nothing. Life restructures material and diverts the flow of energy, carrying elements and opportunity into new places and new stories. On Earth, life has been doing this for the last four billion years, first occupying the oceans and invading the land before leaping into the air on shimmering wings.

Now this eruption has taken on a new form. Space exploration has burst forth from seeds of human scientific thought to inspire a ragtag mix of experimentation and wonder, fueled at times by our age-old appetites for cultural and economic power. Our species has produced ways for living and nonliving things to spread away from the Earth and into the cosmic void above our heads. What were hesitant and fraught orbital trips more than half a century ago have become almost routine practices for the wholesale extraterrestrial transportation of terrestrial matter. Intricate devices are being assembled from planetary elements that originally condensed together from cosmic starstuff, only to now be shot back into space. After the passage of billions of years, it's hard not to see this as a perverse, spitting tantrum that unbinds the stately order of planetary assembly and equilibrium. The world is unmaking itself, and in doing so, it is changing and creating new ecosystems: whether among the tenuous atoms of low orbit around the Earth or with the shells of spent spacecraft that wind up on the surfaces of other cosmic bodies.

These blooms of spaceborne material have often owed their origins to the desires of certain nations and individuals who found themselves sitting atop immense resources from a legacy of dirty conquests and oppressions. But that situation has seen extraordinary change, and launches to space are no longer just the province of the historically wealthiest and most powerful countries. Altogether, eleven nations or joint programs can currently claim to have the raw ability to send things to space,[4] and more than forty-eight countries have had their citizens go into orbit.[5] Today, private space industries regularly supply anyone with sufficient funds the ability to place satellites around the Earth, leading to a tally of eighty nations that have at some point operated their own orbital hardware.

The urge for space exploration has seeped much farther than many of us might arrogantly imagine. Nations such as Nigeria have established their own space programs and have even committed to efforts in the study and practice of space law.[6] The Republic of India first sent a rocket to space in 1963. Today, the Indian Space Research Organisation—the country's space agency—has one of the world's most reliable launch systems and has sent successful scientific missions to the Moon and to Mars.[7] In August 2023, India became the very first nation to successfully land a robotic probe and rover near the lunar south pole. India also has an astonishing number of "space tech" start-ups, with some 140 registered for business and vying for a slice of a global market that anticipates tens of thousands of commercial launches in the coming decade.[8]

Where there are blemishes in the story of our presence in space, they mostly come from the tightly entwined history of certain nations' war machinery with the early space industry, and the grim horrors of human abuse and material destruction in such times.[9] Some historians have gone so far as to argue that war is the sole reason that our species first developed powerful rocketry and spacecraft, although that is perhaps an oversimplification. Today, we're watching the emergence of billionaire technologists and industrialists who exploit—whether knowingly or inadvertently—the enormous romance of cosmic exploration to carve out their own profitable vision of the human future. What history will say of this is yet to be written.

Space, like any pristine environment, can also be polluted to an extent where its most desirable qualities degrade. We're throwing so many objects into orbit around the Earth (there are nearly eight thousand operational satellites as I write this) that our view of the universe is becoming contaminated by their myriad points of reflected sunlight as these machines skim silently across our nighttime views anywhere on the planet.[10] This can thwart the potential of modern astronomical observatories, and it steals away some of our access to the purest kind of wilderness that every human being should expect to be able to explore for free: the compelling starry sky.[11] Even the destructive, fiery reentry of satellites now appears to be a factor in the pollution of the upper atmosphere, with implications for Earth's climate, as I'll discuss. Overcrowding in orbit also has the potential to curtail the very activities that are of commercial value in space by filling Earth's orbital terrain to an extent where junk poses an extreme hazard to satellites.

At an even more fundamental level, every single rocket we launch into space creates the possibility of unforeseeable consequences that could echo far into the future. We, as representatives of Earth's entire lineage of living things, have breached the unseen wall between life on a single world and the rest of the universe. That breach takes parts of an old, deeply evolved planet and its offspring—in microbes, machines, and humans—and disperses them into alien environments completely outside their evolutionary crucible. Whether trapped in endless orbits or marooned on the surfaces of other worlds, these entities are subjected, even if sometimes only briefly, to new pressures and imperatives. This is a cosmic experiment in life's evolutionary trajectory without precedent for our solar system.

The influence of this dispersal is also not one-directional, not just pushing outward. We're still firmly connected to these otherwise orphaned devices and ships: tracking them, listening to them, decoding the discoveries they make, and using that knowledge to tinker with our designs and strategies to make even more rockets and probes to be flung into space. We have become an obsessive receiving channel for information unobtainable by any surface-bound, single-planet species. No other organisms on the Earth have access to this kind of knowledge about not just their immediate environment but also the environment over the horizon and on the other side of the planet, as well as what is happening across the solar system and the observable universe. That's an extraordinarily privileged and influential status.

Detailed sensory data about the Earth obtained from space has been transformative for us and for all the species we've co-opted into our style of life, whether they are now inbred and hybridized food sources or marginalized wildlife. For example, in 1960 the very first functioning weather satellite triggered a profound change in how we carry out all of our businesses of farming and traveling. Today, as I'll show in this book, the human world simply doesn't function without its feed of information from above our heads. We're enveloped by a constellation of more than a thousand orbital monitoring stations designed explicitly for tracking weather and other conditions. These stations also capture the detailed properties of the land, the oceans, and the atmosphere, helping us spot events like hazardous wildfires and measuring everyone's pollutants. Imagine if the Roman Empire or the vast Qing dynasty had been given access to such knowledge. Its decisions and trajectories would have been very different, just as our present course would be entirely altered without these tools.

We've also traveled far enough into the universe to be able to look back and see the entire Earth as a lonely island in space. In the late 1960s and early 1970s, the Apollo astronauts snapped several galleries' worth of color images of the "blue marble" of planet Earth on their journeys to the Moon.[12] Those stunning, iconic images came at a time of rising environmental awareness and caught our collective imagination with their clear and uncomplicated expression of Earth's beauty and unity. Today's annual Earth Day is part of the legacy of that moment and of the ongoing struggle to find a sustainable pathway for our overconsuming civilization.

The modifications induced in human behavior by our cosmic experiences then modify the environments and circumstances experienced by all species of life on Earth, whether through climate states or food chains. We may now be starting to do the same beyond our planetary shores, even learning how to significantly and deliberately modify the orbits of celestial bodies themselves—a strange and unsettling skill unlocked in 2022 when a "moonlet" of an asteroid was deliberately hit by a spacecraft and slowed in its orbit. This was a test of a methodology for protecting the Earth and all of its life from the impact of hazardous asteroids. But it was also a taste of a possible—maybe even probable—future in which the fundamental architecture of our solar system becomes far more malleable than we might now imagine.

It is also no secret that our species is living in the midst of a major evolutionary tipping point. We're faced with terrible uncertainty about the human trajectory, brought about by global warming and resource use. We're also faced with the compounding effects of an ongoing mass extinction all around us that changes the diversity and stability of the planetary biosphere. The revolution of space exploration and the increasingly interplanetary nature of life around this one lonely star we call the Sun aren't often seen as factors in this evolutionary moment, yet they are of the utmost importance. They mark a fundamental change in the possibilities afforded life and the option of a trajectory that sees our species, and others, as both vastly grown and diluted, as well as irrevocably altered.

I call this trajectory into the future "Dispersal." It describes a scenario that is the synthesis of everything that we've learned about the physical principles behind space exploration, the history of that exploration, and the history of how our ideas have changed the nature of our own evolution. Dispersal is an acknowledgment that the sheer scale and diversity of our solar system create a new paradigm for life's future. Most importantly, Dispersal is not a philosophy that anyone has to buy into for it to take place, although things might be easier if we did. Instead, Dispersal seems likely to be nigh-on inevitable. I'd argue that it is a future that life will now be driven toward by the central, universal logic of natural selection and evolution. In fact, in a very real sense the dispersal and alteration of life *is* life. It is an essential component of what differentiates living things from nonliving things.

To fully unpack this and to see what Dispersal is—to create this guiding synthesis of knowledge—we have to understand the cojoined and complicated histories of the worlds of our solar system and the intricacies and novelties of our forays into the surrounding cosmos during the past century. At the same time, those fine-tooth details have to be balanced against a much larger vision of what the phenomenon of life really is and how the nature of physical reality sculpts and is sculpted by that phenomenon.

That's all very grand stuff, but the reality is that most of us spend very little time worrying about space exploration. In this era of informational overload, what's going on far above our heads has a lot of competition for attention. The only exception is when something goes so explosively wrong or so exceedingly right that our news feeds and social media light up for a flashing instant. If we do ever pause to think about the topic, it can easily seem that space travel has not progressed so very much since the 1960s and 1970s, that the age of space is a little bit stalled or mundane. In other words, space falls into the pop-culture category of "Where are the flying cars I was promised?"

Yet the truth is that there has been an extraordinary and accumulating advancement and expansion of space exploration of such complexity and scale that it needs many more than a few sentences to unpack. I'd further argue that the drivers and qualities of this expansion aren't just about human behavior but are also tied to deep principles that govern the physical universe: principles of motion, energy, and gravity that our hominin brains have expended thousands of years toying with and grasping for before finally decoding them and recognizing what we can do with that knowledge.

None of these advances make space exploration or the potential settlement of humans beyond the Earth easy. In fact, this is arguably the most difficult and dangerous endeavor that humans have arrived at in our brief span as a species. It is right up there with "tease the mammoth," "dive into this watery cave," and "try to create a utopian society." No aspect of it is simple. The engineering needed just to get beyond Earth's 100-kilometer skin of atmosphere is extraordinarily tricky.[13] The current prospects for maintaining life in space or on a world like Mars for any real length of time appear, at first glance, fantastically awful. The challenges run the gamut from how to handle our most basic physiological needs in space to the unknowns of how to create functional and equitable societies and systems of governance in alien environments where entirely new hazards and resources become fresh items to squabble over.[14]

Yet . . . all of these hurdles almost certainly appear disproportionately intimidating precisely because we're the very first life on this world to move beyond its confines. We're the pioneer species flowing into a new landscape and making all the mistakes. Consequently, the present time is a moment when abrupt changes to our philosophical stances on the questions of existence really do matter: whether, for example, we think space exploration is a part of nature or separate from nature, and whether we have all been just staring at the shadows of the world, like Plato's unfortunate cave dwellers, before being thrust out into the real cosmos.

An unexpected starting point in coming to grips with this odyssey—its past, its present, and, most of all, its future—is a tale that is simultaneously allegorical and intimately, definitively connected to the cosmic arc of things because it combines all of the exquisite possibilities of life, people, machines, and revolutionary ideas into one singular launch.

On Monday, October 24, 1831, a rather tall, round-faced twenty-two-year-old human by the name of Charles Robert Darwin tumbled from his horse-drawn carriage into the bustling naval community of Devonport (formerly Plymouth Dock), on the southern coast of England. The enterprise he was about to engage in would be a wholly new and extraordinary experience for him, and highly unusual for anyone at the time. It would be an experience almost infinitely removed from his preceding years as an affluent and happy student at the University of Cambridge, in the flat lowlands of eastern England. Although the young Darwin was already a lively and gifted communicator, his journals convey little of his deepest feelings in Devonport. We can only speculate about exactly what was going through his mind as he stood there watching his luggage being unloaded. However, it's a pretty good bet that as much as he expressed his appreciation of the coming adventure in his diary, he was probably also reasonably terrified.

In the docks at Devonport the sailing ship HMS *Beagle* was getting fitted and supplied in preparation for a planetary journey, a literal voyage around the world, to parts largely unknown to many of its crew.[15] For the British government and the royal empire, the *Beagle* was to undertake a critical mission to survey the eastern shores of South America and possibly beyond. The information gathered would help ensure superiority in navigation and solidify knowledge of lands that already supplied (or might supply) the British imperial power with raw materials and goods. For the captain, Robert Fitzroy—a wealthy and hugely capable but occasionally volatile career navy officer—this was a second voyage on the *Beagle* and a relished opportunity to develop his own interests in geology and natural history. For Darwin, offered the role of ship's naturalist through connections with his favorite Cambridge botany professor and some tactical twisting of his father's arm by an uncle, it was the scientific chance of a lifetime. For the crew, it was a skilled and secure yet extremely hard job that they expected to last for about two years.

The *Beagle* itself was a heavily hybridized and customized machine that was built out of centuries of hard-won maritime experience and the expectations of what Earth's oceans could confront a sailing ship with. It had begun life a decade earlier as a vessel of warfare, designed and built as a ten-gun, two-mast brig-sloop that was later modified to three masts. But by the time Darwin boarded, it had been substantially upgraded and had already spent several years proving its worth in the southernmost parts of South America. The *Beagle* weighed in at 242 tons, and its upper deck had been rebuilt and raised by nearly a foot under Captain Fitzroy's directions in order to help reduce the amount of water it would take on in rough seas and to actually make it less top-heavy in those situations. The oak hull was newly and heavily reinforced, and it was equipped with a recently invented lightning conductor. This consisted of metal plates fixed all the way from the masts down to the hull's copper sheathing and connected to all of the major metallic structures on the vessel. The *Beagle* was also, in turn, a mother ship that carried several other boats: a 24-foot sailing yawl with a smaller cutter stored inside it, as well as four open-deck whaleboats and a dinghy.

Despite all of these features, by modern standards this ship was astonishingly compact and claustrophobic: only 90 feet long and 25 feet across at its widest. Crammed into those dimensions was a crew of seventy-four people. The hold was only 13 feet deep at its maximum, so in some places the inner decks were about 63 inches apart, which must have been incredibly uncomfortable for someone like Darwin, who was close to six feet tall. This was, in so many ways, a vessel that had more in common with the cramped environments of today's crewed spacecraft than any large ship that civilians might now find themselves on.

Because of its financing from the British Empire, a lot was riding on the *Beagle*'s capabilities. Accurate timekeeping was essential for navigating and surveying, and this enabled the ship to determine its longitude by comparing the time—locked to a standard datum—to the position of celestial objects in the sky. But the art of engineering precision timekeeping devices for use at sea was still evolving, and under Fitzroy's command the *Beagle* carried no fewer than twenty-two new chronometers to be tested and compared throughout the voyage. This was a subtle but profound evolution in the basic structural approach to handling redundancy and error correction in critical systems, and one that would continue to develop and eventually be implemented in spaceflight. Indeed, Fitzroy's single-minded concerns over his equipment would make modern rocket engineers proud. He even used his own money to replace the iron cannons with ones made of brass to reduce the chances of unwanted magnetic influences on the ship's compasses.

The crew members of the *Beagle* were also carefully and strategically picked. Sixty-five of them were regular, experienced sailors who were supported by carpenters, surgeons, cooks, surveyors, and a cohort of Royal Marines. The rest were nine specialists, including the keen young Cambridge graduate Charles Darwin as well as the artist Augustus Earle. Very little was left to chance in who was selected because once they left the oceanic orbit of the British Isles, there would be strange new environments and dangers that would test all but those best equipped to be adaptable and resourceful.

Like the ships and spaceships that would come after it, the launch of a vessel like the *Beagle* consisted of a laborious number of steps until all conditions favored its moment of release to the oceans. After a month of initial preparations, on November 23 the *Beagle* finally untied from its dock at Devonport and sailed a short distance to a staging post across the neck of Plymouth Sound, where the fresh water of the River Tamar intermingles with the salty flow of the English Channel. Here, the *Beagle* dropped anchor by the leafy Tudor estate of Mount Edgcumbe, at a spot on the water within sight of a monument called Milton's Temple.[16] This small domed folly had been built just above the shoreline in 1755 and to this day is adorned with a plaque carrying a quotation from Milton's epic poem *Paradise Lost* that evokes a landscape at the border of the Garden of Eden: "A sylvan scene, and as the ranks ascend / Shade above shade, a woodie Theatre / Of stateliest view."

Despite this mythological pretension, to the casual observer, and most probably to the bulk of the *Beagle*'s working crew, it must have seemed like a pleasant but forgettable location as a part of the ship's slow progress to readiness. But on the grander geological scale of deep time and evolution, Milton's paradise resonates a little more. The entire British Isles actually tilts in a southeastern direction down to this lush shoreline, the whole landmass still rebounding from the weight of an ice sheet more than one kilometer thick that sat to the north some thirty thousand years before. It's as if the Earth's crust is determined to send things rolling off into the ocean just south of Mount Edgcumbe, the force of gravity willing explorers to push away from their sheltered sylvan scene and launch to another realm.

In this case, as much as Darwin might have wanted to move on and begin the voyage, he and the crew were held in suspense as they waited for favorable winds, giving him time to learn that his body didn't appreciate the bobbing motion of the ship or its finicky sleeping hammocks. To add injury to the insults of seasickness, the *Beagle* then made two failed attempts to sail out of Plymouth Sound just as a series of gales battered southern England that winter. The first, on December 9, was a wholly embarrassing failure to launch, with the ship slinking back to anchor the very next day. At the second attempt, on December 21, they got within sight of Lizard Point, at the southernmost tip of England, before temporarily running aground on a rock and having to retreat once again as storms came in.

By Christmas, it must have seemed to Darwin that they'd never be under way, and while he dutifully—and perhaps gratefully—went to attend holiday services at a church on the firm soil of the mainland, the crew dutifully got into a drunk and boisterous state, so much so that when perfect sailing conditions appeared on the next day, they went nowhere, except to their hammocks to nurse hangovers and punishments for misbehaving. Finally, as December 27 rolled around, a sobered-up *Beagle* made a successful launch and was on its way. This clearly caught Darwin by surprise because he had to race to catch up by riding in a friend's boat, getting aboard just as the vessel entered the ocean proper.

In the end, Darwin wrote in his diary of these two months as "the most miserable which I ever spent." It was a thoroughly inauspicious start to a journey that has now become enshrined in the history of scientific discovery. In retrospect, though, the launch of the *Beagle* was remarkably similar to the painful and glitch-ridden processes that would eventually apply to early spaceflight, a little more than 125 years later.

Technologically supported exploration is never going to be easy because it involves so many variables that are uncontrollable: from the natural environment to the microscopic structural properties of vehicle materials, as well as the complexities of sensory apparatuses. This type of exploration consists of an assemblage of processes and structures that bend and react to what the external world throws at them, sometimes breaking altogether. For the *Beagle*, humans were an integral part of that assemblage. The mind and body of crew member Charles Darwin were instruments of detection and deduction that would wade through tropical waters, jungles, mountainous terrains, rivers and lakes, forests and scrubland. He would experience different human cultures with empathy (or sometimes without), as well as witness radically varied flora and fauna along with everything from violent storms to crystalline night skies to grinding earthquakes amid somber and bleak landscapes where life seemed to be hanging on by a thread. He would see and touch countless rocks and fossils, and be exposed to landscapes sculpted by time measured in millions and billions of years.[17] Soaked in these rich data, he would emerge to interlink the living, the dead, and the structure of the world itself.

In doing so, Darwin himself became an unwitting catalyst of evolutionary change. His new insights about the origins and selective forces of life eventually caught fire in the world of human thought. Darwinian evolution was a framework that would help create tools to transform life just as much as they describe life. By 1944, some eighty-five years after the publication of Darwin's *On the Origins of Species*, the physicist Erwin Schrödinger astutely identified the possibility of heritable biological information being contained in aperiodic molecular structures. By 1959, Rosalind Franklin, Francis Crick, James Watson, and Maurice Wilkins had done the critical work to identify exactly what those structures looked like—in the double-stranded helical polymers of DNA. Today, we have tools like CRISPR-Cas9, expressly designed to edit the genes encoded in DNA in order to fix dysfunction or create new function, but also to divert the flow of natural selection.

Along the way, we've applied the ideas of evolutionary theory to many things in the world. In some instances, we've directly influenced the genetic inheritance of species with refined methods of hybridization.[18] In other arenas, we simply developed strategies of behavior or commerce that produce a cascading effect down the lineage of life, whether in our farmed monocultures or in selective pressures brought about by attempts to eradicate or nullify pests and pathogens.

We've adopted evolutionary concepts into economics, where we talk about adaptation and survival of the fittest, as well as the role of psychological evolution. Simplified evolutionary mechanisms of variation and selection are incorporated into computer science and engineering. We build "genetic algorithms" that perform optimization tasks or zero in on algorithmic solutions to problems largely without human supervision, relying instead on the same weighing of success and failure that applies across the biological world. Building from these ideas, today's astonishing advances in artificial intelligence and machine learning have their heritage deep in the concepts of feedback-induced evolutionary change and adaptation.

All these things have drastically affected our species in the past century and have consequently affected all species of life on the planet. Darwin and his contemporaries didn't just produce ideas about how the world changes; they also produced ideas that caused new kinds of change in the world. In a real sense, the evolution in humans and human tools that helped create and launch the *Beagle* on its journey also provided the means for us to learn enough about evolution to turn around and change the process of evolution. It resulted in nothing less than the evolution of evolution—it was a point of inflection in the history of living things. This is why the voyage of the *Beagle* has more than a passing similarity to the complicated history and future of space exploration. Both that wooden ship and the ships of space involve more than inspirational ideas; they also provoke tangible changes in life on Earth and in the possible scope and nature of those changes.

In the simplest biological terms, whenever life finds its way into new terrain, the potent mechanisms of variation and natural selection will start to discover new solutions to the problem of existence. When that new terrain extends onto scales of space, time, and matter that are millions of times larger than they have been, we have to anticipate solutions that are shocking in their inventiveness and breadth. Dispersal is one such solution.

Since Darwin's time, scientific research has also unpacked a great deal more of the messy details that operate within the grand boundaries of evolution. Organisms are molded by their environments, but they also mold those environments across all scales, whether in the concentrations of biochemicals contained within microliters of fluids or in global cycles of nutrients and energy. The popular idea of genes as direct blueprints for biological function and adaptation, or as irreducible units of natural selection, now seems less and less accurate. It may be better to think of genes as malleable algorithmic seeds that code for interactive mechanisms and functions that then unfurl into ever more complicated systems and are not truly fixed in space or time.

That algorithmic vision certainly fits with our present technological age, where humans don't just alter environments but rebuild them completely, sometimes atom by atom. Space exploration is a preeminent example; it is a process that is entirely dependent on augmentations to living systems. Those augmentations may be in the form of speeding rockets and communicative, autonomous robots or in the form of encapsulated habitats to sustain life. These technological components have their own forms of natural selection and evolution that are sometimes intriguingly similar to aspects of biological evolution.[19] For example, we've learned that microbial life plays merry havoc with comforting ideas of inheritance by engaging in horizontal gene transfer. Single-celled bacteria regularly exchange snippets of genetic material with species that may be only remotely related, scrambling any orderly "tree of life" but gaining functions and opportunities. Technology is also rife with shared and stolen ideas, sometimes obscuring where these originated but enabling incredible flexibility in the face of new terrain.

Take all of these phenomena together, and there is good reason to believe that a species that develops the capacity to launch itself, and other life, beyond its planetary point of origin really does pass through a new kind of evolutionary bottleneck. It indicates the possibility of a monumental transition. Writers and thinkers have long appreciated that space travel could mark a major change in the history of our species, and they have often made grandiose predictions and claims about the implications for the future. But they, and the rest of us, have not always recognized the intricate and powerful interplays between the acts involved in space exploration and the evolutionary shifts that are already being forced on humans and on the rest of life on Earth. Understanding those existing interplays could help improve our predictions for what is yet to come. As with Darwin's colorful story, we're not going to be able to fully understand the future implications of space exploration without carefully scrutinizing its history and present trajectory.

Just as the extraordinary voyage of the *Beagle* helped launch a new theory of evolution that unlocked new evolutionary opportunities, the present voyage off the planet is set to do the same. And if we're smart enough, we might just find a way in these opportunities to improve our chances in the cosmic night. This is the story of how that might happen.

— 2 —

A HISTORY OF
PROPULSIVE IDEAS

In the case of any bird soaring, its motion must be sufficiently rapid so that the action of the inclined surface of
its body on the atmosphere may counterbalance its gravity.

—Charles Darwin,
The Voyage of the Beagle

Space exploration may be a supremely important evolutionary
turning point, but there is no denying that life on Earth has taken
nearly four billion years to begin to lift itself into the cosmos. This
raises the simple but profound question of why this hasn't happened
before now. A part of the answer lies in the many contingencies of
evolution, where a mixture of small and sometimes large consecutive
steps are necessary in order to unlock new possibilities in the future.
In the case of space, those steps were all about very special ideas that
had to be formed and acted upon. Both the voyage of the *Beagle* and
space exploration have involved stepwise developments in the fundamentals of how we model the physical world and in how those models

manifest themselves in our actions. For example, there is celestial navigation, which uses observations and models of the movement of stars and planets, and there are models and maps of planetary phenomena like the Earth's rotation and its magnetic fields that dictate a need for a vessel like the *Beagle* to be so well-equipped with chronometers and compasses. But there are also critically important concepts for our exploration of the world that are much more esoteric and abstract—like those of gravity and the theories of energy and its many manifestations.

Where space exploration diverges from a maritime voyage is that reaching space would be almost inconceivable without the ecosystem of mathematics, physics, and chemistry that has developed across the past couple of thousand years in multiple cultures and circumstances. It might be possible to build an explosively powerful rocket to launch away from the Earth with only an empirical, artisanal knowledge of how the world works—gained by constant trial and error—but it seems almost impossible that this could ever provide the precision and reliability that is so essential for genuine exploration. Similarly, the challenge of negotiating the complexities of orbital mechanics to move and transport spacecraft around the solar system would be extraordinarily difficult without a theoretical basis in mathematical physics that, though not easy to implement, can provide immutable truths about what you can and simply can't do.

Life's long lead time before escaping the Earth is also a consequence of the fact that reaching space involves overcoming a very special kind of physical barrier for organisms that sit in the filth of a planetary surface. This barrier is different from the intimidating expanse of a great ocean, the rugged mountains and deserts of a continent, or even the variations and currents of an atmosphere. Those environments all represent major challenges of isolation and scale, and can test physiological maintenance to its breaking point in conditions that range from hot to cold and wet to dry. But to simply get into those environments is not so difficult: life can often just scatter or ooze its way through. For space, though, nothing matters unless you can first overcome the fundamental problem of gravity.

Starting from scratch, it's not entirely obvious what mechanisms can be used to climb out of the gravitational well of a planet. An organism could climb in a literal sense by just building a high enough ladder or staircase and walking its way to the universe—a tower of Babel poking into the heavens. That's perhaps not as unfeasible as it sounds (as we'll see), but humans have instead taken a more impatient approach by pushing our way upward with rockets. On the Earth, that involves a number of fundamental choices. For example, you could expend an enormous (and ultimately impractical) amount of continuous propulsive energy to hover farther and farther away from the planetary surface. Or you could expend a burst of energy to propel yourself in one heroic push to a vast distance where the gravitational pull of the planet no longer bothers you very much. Or, if you're clever, you could choose an intermediate option, where you expend significantly less energy to throw yourself into a sustainable orbit.

The catch is that in these latter two cases, the velocities that have to be attained are huge compared to anything that life on Earth usually confronts. To fling yourself away from the Earth to never return means reaching what is termed an *escape velocity* that amounts to 11.2 kilometers *per second* (or 40,270 kilometers per hour) from the planet's surface.[1] To reach a sustainable orbit a few hundred kilometers above the Earth involves reaching a velocity of some 8 kilometers per second, a rate that would carry you the entire length of Manhattan Island in about 2.5 seconds. To loft a human of average mass into that orbit takes as much energy as is stored in about 1,200 fully charged modern electric cars.

Inside the Earth's atmosphere, the density and friction of the surrounding gases make velocities like these essentially unmanageable for common machinery. In other words, to get to these high speeds you really have to get out of the atmosphere as soon as possible. Consequently, rocketry boils down to moments of high acceleration and terror, where almost any deviation from the intent of propulsive mechanics will produce a null result or, worse, complete disaster. There are not many physical actions in the world of biological evolution that hang on quite such a thread, where the future is entirely predicated on the perfect functioning of a briefly unstoppable surge of so much energy.

That energetic cost points to a more sophisticated response to the original question of why it has taken billions of years for life on Earth to purposefully lift itself into space. Measured in cosmic terms, the phenomenon that we call biochemistry is extraordinarily thrifty when it comes to energy, and natural selection abhors inefficiency. The kind of *density* of energy needed to form complex molecules, string them together into polymers and proteins, or even to coordinate great clumps of them into oozing and flexing, perambulating masses is very modest compared to many other phenomena in the cosmos. The energy density in a biological cell is tiny compared to the cumulative energy density of the nuclear furnaces of shining stars or even the low, grumbling heat deep down in a molten planetary interior.[2]

Estimates of the total power flowing through Earth's entire biosphere are not easy to make, but they range from around 100 terawatts (a terawatt is 10^{12} watts) for all active metabolic biochemistry to as much as 5,000 terawatts, or 5 petawatts (a petawatt is a quadrillion watts, or 10^{15} watts).[3] That larger upper estimate includes biophysical phenomena like plant evapotranspiration, which lifts a global volume of more than 70,000 cubic kilometers of water from the ground up through the vascular system of plants, transporting nutrients, and eventually evaporating into the atmosphere, forming a key piece of Earth's water cycle.

Superficially, these estimates sound like an awful lot of power, and they certainly dwarf humanity's present technological power use, which registers at about 15 terawatts.[4] But the total amount of solar power that the Earth's atmosphere and surfaces absorb is around 122 petawatts. The Sun's total output is around 100 *billion* petawatts, and even the planet Jupiter—a goliath of brooding hydrogen and helium more than three hundred times the mass of Earth—glows with an internal flow of some 460 petawatts. By comparison, life on Earth is barely sipping at what is laid out before it, like the most painfully polite dinner guest. Single-celled microbes, like the bacteria that are quietly wedged into the nooks and crannies of Earth's soil and rocks, have individual power budgets somewhere around a billionth of a billionth of a watt. For comparison, an ordinary smartphone being used to browse the web slurps up as much power as a billion billion microbes.

In other words, life has to this point evolved to require far, far less power density (less spatial *concentration* of energy flow) than is needed to fight against gravity's barrier in any substantial way. If anything, biology has steadfastly gone in the other direction than it should have if it needed to escape into the cosmos. For humans, or any other life on Earth, to have a chance of reaching space, there first had to be a revolution comparable in scope to the revolutions of photosynthesis, or of multicellularity and nervous systems, except that this revolution would take place in a different fabric, the fabric of ideas. In a very real sense, the ultimate evolutionary trait for overcoming gravity's barrier was thought itself.

The story of that revolution takes us to the heart of what we even *mean* when we talk about this mysterious thing called energy or when we talk about overcoming an equally mysterious thing called gravity, and about flinging ourselves into something called an orbit. It's an epic tale stemming from all the human cultures that have ruminated on the nature of the heavens. Those origins also owe a debt to every human child who has thrown objects and learned the puzzling disappointment of how they fall down but not up. A debt is also owed to great thinkers across the planet, from the Mediterranean civilizations to the Islamic world to the Indian subcontinent and beyond, who agonized over explanations for these observations long before we even had names for the phenomena that we now call mass, motion, and gravity.

As luck would have it, the greatest steps in the story of how life has reached space came from connecting the threads between observations of the configurations of the planets and stars, and the revelation that there are universal laws of nature. Finding some of those laws was intimately linked to the issue of heliocentrism—the once on-again-off-again and often blasphemous idea (in at least some parts of the world) that the Earth is not the center of the solar system or of creation itself. The displacement of the Earth from the center of existence was a big deal, but an equally big deal was finding a reason *why* this world and all the other planets in our solar system are constantly moving through the universe the way they are. It was a conundrum that many thinkers wrestled with, but the person who did more than anyone else to finally fit those puzzle pieces together, just so, was Isaac Newton in the seventeenth century, and he did this by explaining how a set of *rules* of motion and forces is at the heart of almost all things.

But connecting the dots between that revolution in thought and today's space exploration is complicated, and it is a history most easily told selectively, through the story of a few people at the right place at the right time and with the right connections.[5] In this case, that story starts with the lives of three intellectually dazzling women whose works spanned three centuries.

The first of these people was Émilie du Châtelet, who was no average human—whether measured by circumstances or by personality and intellect.[6] Born in Paris in December 1706, she was the daughter of one of the chiefs of protocol in the Versailles palace of French king Louis XIV, also known as the Sun King. Despite her family status in this rarefied and indulgent layer of the upper, upper classes of France, it was still an environment that didn't particularly value the education of women any more than did the rest of society. Du Châtelet was a force of nature, though, and with parental backing she contrived to study and to gain an entry into the philosophical, scientific, and mathematical community of the time. The only thing that apparently matched her lust for intellectual adventure and growth was a lust for good times at high-stakes gambling tables, as well as quite a few amorous relationships.

At the time of her tragically early death at age forty-three, from complications associated with the birth of her last child, she had become a fixture among the scholarly elite. Not only was she notorious for her long-term extramarital relationship with the multitalented French writer and scholar Voltaire, but du Châtelet was also very much her own intellectual star. She had a capacity to stir and influence the most renowned thinkers of the age, a cohort of minds that appears to have included the philosopher Immanuel Kant and his ideas about the continuity of space and time. In this mix of people, du Châtelet presented as an intellectual sparring partner, a keen questioner, and an academic producer of substance in her own right. Her most widely celebrated and influential contribution was an immensely skillful translation of Newton's monumental mathematical treatise, the 1687 *Principia* (all three volumes of it), from Latin into French.[7] Her commentary in those pages proved essential for European scholars who were still coming to grips with what was often perceived to be a tricky and intimidating text.

For all of these reasons, du Châtelet found herself deep in the intellectual fray as people navigated the complicated implications and subtleties of the revolutionary scientific insights of the previous century. Although the framework provided by Newtonian physics was remarkable, it still contained some surprisingly ill-defined or confusing concepts and gaps. This was a condition that also applied to the works of his contemporaries and archrivals such as the German polymath Gottfried Wilhelm Leibniz—who independently developed the mathematics of calculus. For space exploration, as it would roll into view two centuries later, some of du Châtelet's most essential contributions revolved around her deeply insightful and influential discourse on some of those gaps and on the nature of what we now call energy.[8]

Energy is a peculiar concept. It is not a "thing" in and of itself, despite the many earnest science-fiction stories and movies where mysterious beings "made of pure energy" crop up with dismaying regularity. Energy is a quality that we can assign to the contents of the world in order to understand what is happening, how it happens, and to a large extent why. The earlier Greek philosopher Aristotle thought about the innate properties of matter somewhat differently from the way science does now, but he considered the "potency" of objects and their capacity to do things. In modern physics, energy is the quality associated with matter and radiation that can give rise to actions, to having things happen. Energy helps us calculate with exquisite accuracy and precision how change takes place in the world, whether it is a ball rolling down a hill, a pot of water coming to a boil, or a pair of subatomic particles evaporating into a puff of electromagnetic radiation.

We can create inventories of energy in objects and systems that even encompass the entire universe itself, where the changing fabric of space is hostage to opposing forces in gravity and the bubbling quantum pressures of an empty vacuum. The accountancy of energy allows us to not only predict the future outcomes of physical situations and understand the complex hierarchies of systems that extend from nuclear physics to biology; it also provides some explanation of why stuff happens the way that it does.

Yet until people like du Châtelet started coming to grips with a universal picture of energy, even the notion of *inventing* such a concept was challenging because it was a property of the world that lurked like a blurry figure out of the corner of your eye. To add to the confusion, there is another vital property of the world called *momentum* that shares some of the same qualities. Momentum is the mathematical product of an object's mass and velocity, and by the mid-1700s, Newton's laws of motion had codified the observable fact that momentum tends to persist, even if objects bounce around off other objects. In fact, if you take a system like a set of perfectly bouncy balls (the technical term is actually "elastic" balls), the total momentum—the sum of all those masses times their velocities—remains exactly the same as time passes, no matter how many collisions and bounces take place. In other words, the total momentum tends to be *conserved* for objects and for systems of objects as long as other secondary effects like friction are small.

Although we experience this all the time (it's part of why a batter can send a baseball into the stands by exchanging and conserving the total momentum of the bat and the ball), it's a deeply strange property of the universe. As Newton articulated in his first and second laws of motion, anything at rest or moving in a straight line will continue to do so forever unless acted on by a force. And when this force acts on an object, it causes the momentum to change at a specific rate that is equal to this force (or, to say this another way, acceleration is force per unit of mass). This implies that an object's motion is a property of that object in the same way that it may have a color, a shape, or an electrical charge, and it will change only when something happens to it. In the English language, at least, we don't really express the motion (or lack of motion) of things as an intrinsic property; it's more like something that happens to objects. But for physics, in any given instant that intrinsic nature is exactly how motion and momentum are treated, which can seem pretty weird.

Perhaps that weirdness contributed to the lengthy gap in time between Newton's formalization of the laws of motion (that included momentum) and the rigorous development of the concept of energy. Perhaps it was also the fact that a lot of the intellectual elite, especially in Europe, just didn't agree with Newton's deference to the truth of the mathematics behind his proposals, where a force like gravity could reach mysteriously across empty space. Instead, some other scholars still hoped for something more explicable within the framework of religion and philosophy of the time.

For these reasons, before the time of thinkers like du Châtelet, most natural philosophers were rather vague about the idea of energy, and they considered it as another kind of property that was more or less equivalent to momentum and that seemed to be conserved in systems. Newton himself thought that energy, just like momentum, was a quantity directly proportional to an object's velocity. But du Châtelet, a voracious reader, was intimately familiar with Leibniz's work, and Leibniz had not agreed with all of Newton's theories.[9] Instead, Leibniz had proposed an alternate theory of motion that made use of two new concepts: kinetic energy and potential energy (although the latter wasn't labeled this way at the time). Leibniz saw—correctly as it turned out—that Newton's conceptions of momentum and energy didn't quite complete the physical description of how objects behaved in the world. Leibniz considered kinetic energy to be the vis viva ("living force") of matter and saw it as a quantity that would also be conserved. He further brought up the idea of a "motive force" that was equal to the product of an object's weight and its height above the ground. For Leibniz, this motive force, in effect, represented the amount of stored change or action that an object could bring to the world. After all, if you raise an object up and then release it, the higher up it was, the faster its eventual impact on the ground. Today, we call this stored action the gravitational potential energy. Critically, in Leibniz's formulation the kinetic energy (the vis viva) was equal to the mass of an object multiplied by the *square* of its velocity, and for a falling object this could, in turn, be related to its original gravitational potential energy.

But his particular use of all these terms like *motive force* and vis viva was, frankly, a bit of a mess and was definitely unfortunate:

Leibniz's kinetic-energy formula represents one of the great "almost got it right" moments in physics. We now know, after many more investigations by physicists in the nineteenth century, that the true formula for kinetic energy is simply Leibniz's calculation of mass times the square of velocity, but divided by two. That may not seem like such a big deal, but that factor of two is the difference between experimental measurements making sense or being nonsense, and the difference between reaching space or dying in a catastrophic rocketry nosedive.

To be fair, during Leibniz's time there was little experimental data to go on to even support his claim that it was the square of velocity that measured kinetic energy. But by the time du Châtelet came on the scene, a Dutch physicist called Willem Jacob 's Gravesande was performing experiments in which heavy balls were dropped from different heights onto soft clay to measure the size of the deformation they caused and to relate these values to kinetic energy and potential energy, as they would later be called.[10]

The story goes that du Châtelet learned of these experiments and even repeated them for herself. From the size of the deformation of the clay's surface, it was possible to show that kinetic energy was indeed likely proportional to the square of the final velocity of the balls (after they accelerated down from being dropped). It was also possible to show that the size of the deformation was directly proportional to the height that they were dropped from, multiplied by the weight. This was intriguing because it provided empirical evidence that one definition of energy—the motive force of Leibniz, the potential energy—was indeed directly related to another definition of energy, the kinetic energy.

Du Châtelet eventually realized that it wasn't just momentum that could be conserved; the sum of these energy qualities might be conserved as well. And if that happened in an ordinary system of dropping balls, then by extension it might be a universal law. In more precise, modern terms, the conserved quantity was the total energy: the summation of potential and kinetic energy at any instant. At the outset, when a ball sat motionless at the top of the experiment, all the energy could be said to be in the form of potential energy. At the last split second when the ball hit the clay, all the energy could be said to be in the form of kinetic energy. In between, as the ball fell, energy was, in effect, moving from its potential form to its kinetic form—it was being exchanged while the total remained constant throughout.

Although the case was far from closed at the time, from her records and writings about her observations of this simple experiment and its implications, together with her role as a sharp intellectual influencer in physics, du Châtelet helped set the stage for the development of some of the most elegant and useful concepts in modern physics. In doing so, she also played her part in setting humanity on a course to the stars, because it is this very same conservation of energy and its exchange between forms, along with the conservation of momentum, that provides not only a full description of the nature of orbital mechanics but also the essential rules of propulsion in rocketry or by any other means.

Before those pieces could come together, though, there were two more chapters in the evolution of ideas that provided critical dovetail joints. The first of these stories involves the Scottish scientist Mary Somerville, who was born in 1780 and worked until her death, in 1872.[11] Like du Châtelet, Somerville was born to a family of some status (her father was an admiral), but unlike du Châtelet, that didn't mean any real wealth. Fortunately though, Somerville's parents made sure that she had a decent education, and after her first husband died when she was just twenty-seven, she found herself with resources to pursue her interests in mathematics and astronomy. Within just a few years, she had built a reputation for solving difficult mathematical problems and was becoming more and more engaged with the scientific world. In fact, during her life Somerville interacted with an extraordinary range of people and their scientific pursuits. These included the astronomer William Herschel, who discovered the planet Uranus; the polymath Charles Babbage, who originated the concept of digital computation; Michael Faraday, the creator of electric motors and discoverer of many of the principles of electromagnetism; and eventually the great scientist Pierre-Simon Laplace.

In her early thirties, Somerville began studying Laplace's five-volume, highly mathematical "summary" of gravitational physics since Newton, called *Mécanique Céleste (Celestial Mechanics)*.[12] These books contained many of the advances that du Châtelet's scientific colleagues had been hard at work on decades earlier, and they included Laplace's application of calculus, which would eventually become the default approach in the physical sciences. Because of Somerville's expert grasp of Laplace's heady mix of mathematics and physics, she was eventually asked by the Society for Diffusion of Useful Knowledge (an organization devoted to making knowledge accessible to all) to produce a popular account of *Celestial Mechanics*, in 1827, just after Laplace's death.[13]

What she ended up creating was far more than a simplified popular version. Her book was called the *Mechanism of the Heavens*, and it took the first two volumes of Laplace's opus and filled in the mathematical blanks. Somerville skillfully illuminated the places where Laplace had made leaps of insight that left most readers in the dark—and she connected this version of physics directly to Newton's approach. In doing so, she brought clarity to the confusion over Leibniz's vis viva and kinetic energy, as well as one of the most profound elements of physics: the principle of least action. This is a principle that is inextricably entwined with calculus (and the calculus of variations, to be more precise) and was an idea ruminated on by du Châtelet and others. The principle has many, many applications, but it can be represented in simple terms by asking how it is that an object sliding down a surface because of gravity will "find" the quickest path down. It turns out that calculus can allow you to determine that path. In fact, if the surface has a mathematical description, you can find how to minimize the "action" of the sliding object.

In explaining Laplace's analyses, Somerville pinpointed the extraordinary fact that Newton's basic laws of motion and, by extension, the universal rules of behavior under gravitational forces can be derived mathematically from that same principle of least action. To put this another way, the universe seems to abhor wasted effort and consequently seems to select one path or trajectory for objects out of what could be, in principle, an infinite number of ways that things can go from point A to point B.

At this point in history, however, the concepts of energy and its conservation were still frightfully muddy. Somerville, in interpreting Laplace's work, found herself grappling with some of the same puzzles that du Châtelet had faced in trying to figure out the connections between Leibniz's intuitive physics and Newton's mathematical but more obscure rendition of the same ideas. In fact, it wouldn't be until the 1850s and into the 1870s that experimental work by physicists like Lord Kelvin and James Joule produced the exact definitions of energy and work that we have today, as well as confirmed the idea of equivalence between types of energy.

Perhaps Somerville's greatest accomplishment in interpreting Laplace's work, as well as work on the calculus of variations by Joseph-Louis Lagrange (whom we'll meet again later), was her exposition of the theory of *perturbations* as applied to the motion of planets and objects in the solar system.[14] There's a famous part of Newton's investigation of gravity where he states (paraphrased here) that although it's easy to solve the equations that describe two objects orbiting around each other, like the Earth and Sun, to solve exactly the equations describing the motion of three or more objects appears to be beyond the scope of any human. In a memorable example of scientists saying "Hold my beer," Laplace, Lagrange, and others in the 1700s and 1800s came up with a brilliant albeit exceedingly technical way to at least partially solve this problem.

Newton had actually already suggested a way to try to approximate the effects of the many gravitational pulls that might act on an object. In Newton's fix, you took the mathematical description of a simple, idealized two-body orbit and then gauged how the real world might perturb that solution and an object's motion. What Laplace and Lagrange did was to put that idea on calculus steroids, extending it to the solution of the motions of all planets and objects in the entire solar system. They did this by recognizing that the way the planets' gravitational forces perturbed one another's repeating orbits was itself periodic, except that those perturbing periods could span thousands and millions of years before they repeated themselves.

The mathematics behind these ideas is intense, but Somerville provided a usable guide that laid out the tools to help explain the nagging peculiarities that astronomers were measuring in planetary motions. This was, in retrospect, a critical step that prepared science for the revolutions of Albert Einstein's theory of general relativity, which would upgrade and transform Newton's version of gravity. But the increasing rigor of these mathematical devices also helped set the stage for our third player to illuminate an even deeper piece of the nature of the physical world and a path to exploring the cosmos.

By the early twentieth century, the very nature of the conservation of qualities like momentum and energy was itself subject to some extraordinary intellectual insight. Central to these efforts was the mathematician Amalie Emmy Noether, who was born in Germany in 1882, just a decade after Somerville's death and almost exactly 155 years after Newton's death. Although she too struggled with society's prevailing attitudes toward women academics and experienced misogynistically dismal institutional support, by the early 1900s Noether became recognized as a once-in-a-generation mathematician.[15] Albert Einstein would later write of her, in a letter to the *New York Times* in

1935, as "the most significant creative mathematical genius thus far produced since the higher education of women began."

Among her many other contributions to mathematics, in 1918 Noether discovered—using mathematical methods from Lagrange—a profound connection between symmetries in systems and their laws of conservation. What became known as Noether's Theorem explained, in effect, where the conservation of properties in nature came from in the first place. For example, the symmetry of space translation (if objects remain the same when moving or translating back and forth, or by changing the coordinate system) is what gives rise to the conservation of momentum. The symmetry of time translation (if objects and their behaviors are unchanged through different time intervals) actually gives rise to the conservation of energy.

Because these translations in space and time boil down to saying that the laws of physics don't change depending on where or when you determine them, this means that the conservation of qualities like momentum or energy is deeply, profoundly connected to the fabric of reality. Noether's Theorem tells us that if we look hard, we will, with complete certitude, find properties of the world called invariants, which are conserved. In fact, the conservation of energy is ultimately a consequence of the fact that (as far as we know) physics today is the same as physics yesterday and tomorrow.

People like Noether, Somerville, and du Châtelet all helped change the fundamental way that we think about the nature of the world and the universe beyond. Hypotheses about unseen forces like gravity and unseen properties like energy and momentum took centuries to come into proper focus. As they did, though, they sent ripples through many, many other fields, from political philosophy and social philosophy to economics, where people like Locke and Hume adopted the successes and implications of inductive reasoning, which was bound together with a mathematical vision of reality. These perspectives also changed our conceptions of "place" in the cosmos and of the rules that separate the possible from the impossible.

Just as Darwin and his contemporaries brought about one evolution in thought, the peeling apart of the workings of motion, gravity, and energy to find the cosmic in the colloquial was intertwined with great changes in *how* we thought. The history of space exploration, with its roots in these shifting cognitive frames, was never just a thread of mechanical achievements; it was a long-germinating agent in the human psyche.

TRANSLATING ALL OF THESE deep physical insights into methods for actually getting into space and becoming interplanetary is, however, a whole other story. It is one thing to quantify the energy and momentum of planets emplaced in the cosmos, but it is another thing to wander among them. Newton's posthumously published work *A Treatise of the System of the World* (1728) describes a beautifully simple and intuitive thought experiment to explain how you might transition from the surface of the Earth to move in an orbit that never again touches the planet.[16]

Imagine, he says, a projectile (that many have later interpreted to mean a cannonball) fired horizontally from the top of a tall mountain. (See Figure 2.1.) Based on everyday experience, we'd anticipate that the trajectory of this object will curve down until it hits the ground. That curvature, Newton says, is caused by the pull of gravity toward the center of the Earth that accelerates the projectile downward from its horizontal path. The faster you fire the projectile, the farther it will make it before hitting the ground. In principle, then, if you fired it fast enough, its path would curve at precisely the same rate as the Earth's spherical surface and it would never hit the ground

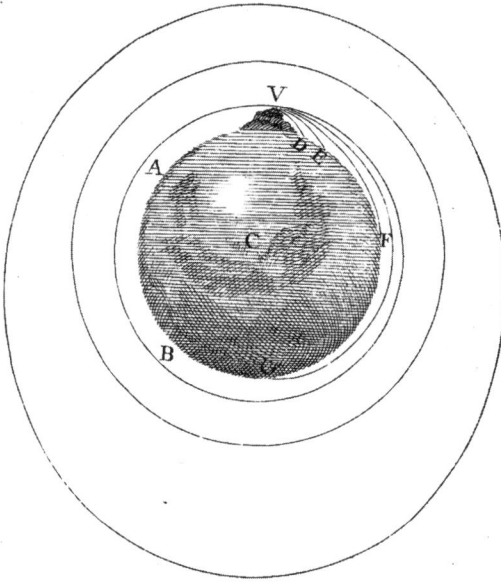

Figure 2.1. Newton's diagram of projectile motion from a mountaintop on the Earth (point V). This version shows several trajectories that eventually hit the surface, as well as a trajectory from the mountain to its antipodal point (G). Faster projectiles will miss the surface of the Earth altogether.

at all! That special path is an orbit. Fire the projectile even faster, said Newton, and the shape of the orbit will change: it will become a larger ellipse with the nearest point to the Earth still at the top of the mountain.

This description contains much of what we need to know about movement in the cosmos. First, orbits are when objects fall—without further propulsion—under the pull of gravity.[17] Anything else, such as when you're busy firing a rocket or accelerating in some other fashion, is not an orbit; it is a maneuver. Because freely falling in a gravitational field means that all parts of an object are accelerating at the same rate, no weight is felt by objects in an orbit. Second, to tweak the size or shape of the orbit, you must affect the velocity, or, in the terminology of rocketry and celestial mechanics, you must apply a "Delta-v," meaning a change in velocity. Finally, Newton's simple example sets the stage for incorporating how the gravitational force weakens the farther away you are from a mass, in proportion to the inverse of the square of the distance.

However, the next step is to figure out exactly how the application of a Delta-v actually modifies an orbit, and that's where there is a direct link to the nature of energy and its conservation that vexed Leibniz, du Châtelet, and many others over the centuries. Newton's mountaintop projectiles help illustrate that closed orbits are, in general, elliptical. In fact, orbits are described mathematically as "conic sections," whose shape can be any of the shapes you find if you slice through a cone with a flat plane. That includes shapes that are not closed circles or ellipses but parabolas or hyperbolas that correspond to open trajectories where an object like a comet may sail around the Sun once only before arcing back to an endless outward journey.

In a closed orbit, the quantities of kinetic energy and potential energy are also held in balance. In fact, twice the kinetic energy summed with the potential energy (that is treated as a negative quantity) is always equal to zero at any point in an elliptical or circular orbit.[18] If the kinetic energy is changed by applying a Delta-v, the orbit must adjust to ensure that this sum remains zero; therefore, the potential energy must change, and consequently the shape of the orbit must change.

Mathematically, this balance can also be expressed in an equation that, rather confusingly, is called the vis-viva equation in homage to Leibniz. The vis-viva equation is key for astrodynamics because it relates five properties: the velocity of an object in orbit, the distance at that instant from the focus of the orbit (the center of a circle or the focus of an ellipse), and the overall size of the orbit. It also incorporates the mass of what's being orbited (a planet or star) and the intrinsic strength of gravity—Newton's fundamental and unchanging gravitational constant. Consequently, you can plug in any values for an object in an orbit and figure out what Delta-v is needed to change to another orbit at any point.

Because of its utility, Delta-v is really the *currency* of space exploration: it is the most directly relevant physical property to talk about when getting around in a place like our solar system, and creating it incurs costs. Whether you want to add a Delta-v to the velocity of your spacecraft or subtract a Delta-v, you have to either fire your propulsion system and expend propellant or perform some other trickery (like using a planetary atmosphere to slow down). The biggest catch is that Delta-v's for space exploration are huge compared to the velocities that life on Earth is used to. Consequently, spacecraft engineers and celestial navigators agonize over the budgets of Delta-v's necessary to explore the solar system. You can even draw up a map of Delta-v's for getting around between destinations that looks like a map of public transport in a large and somewhat poorly designed city.[19]

Imagine that you want to plan a trip to Mars. Your starting Delta-v budget will include what's needed to launch from Earth's surface into orbit, and that might involve a Delta-v of around 8 kilometers per second. Then you decide to follow what's considered close to the most efficient trajectory in terms of balancing Delta-v and time spent in interplanetary space: the so-called Hohmann transfer, named after the German physicist Walter Hohmann, who published the idea in 1925.[20] This involves exiting your orbit around the Earth and putting yourself into an orbit around the Sun. That orbit has to be elliptical in shape so that it will carry you out to the distance of Mars at the orbit's far point (or apoapsis) and, if you get the timing right, arrive at Mars just as it is passing through the same spot as you are in its own path around the Sun. (See Figure 2.2.)

The beauty of this approach is that the Earth is already orbiting around the Sun at about 30 kilometers per second, so you simply need to add on to that velocity. Using the vis-viva equation, you'll find that it will cost a Delta-v of about 3 kilometers per second to enter the Hohmann elliptical orbit around the Sun that intersects with Mars some six to nine months later (depending on the timing). Then, when you arrive at Mars—as it scoots along in its own orbit—you'll need to apply a Delta-v of about 2.5 kilometers per second to match pace. If

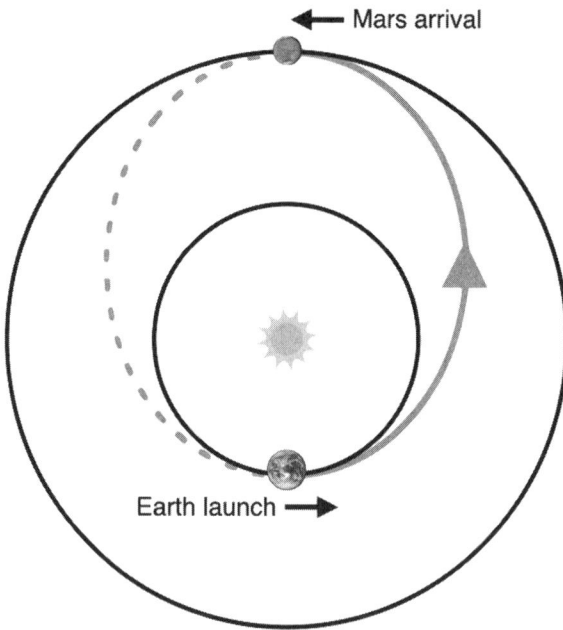

Figure 2.2. Diagram of a Hohmann transfer orbit from Earth to Mars.

you want to get into orbit around Mars itself or go directly to landing on its surface, you'll have to budget for even more Delta-v changes. The upshot of all of this is that any space mission has to have a very precise plan for its maneuvers and Delta-v requirements at every step. That is, in the broadest terms, what space exploration looks like on paper, but it still leaves us with the puzzle of how to create Delta-v and to somehow claw our way off the surface of the Earth in the first place.

In the story of the development of modern rocketry, nobody really surpasses the intellectual contributions made by the Russian scientist Konstantin Tsiolkovsky.[21] Born in 1857 in central Russia, Tsiolkovsky was one of seventeen siblings. At age ten he contracted scarlet fever, which mostly destroyed his hearing and his chances at a formal education. He was extraordinarily precocious, though, and his parents eventually sent him to Moscow, as an older teenager, where he ended up under the wing of the brilliant yet eccentric Russian philosopher Nikolai Fedorovitch Fedorov. Fedorov was an ardent proponent of a school of thought known as Russian Cosmism. This was an all-encompassing, wide-eyed (and rather wild-eyed) view of humans and the cosmos that blended philosophy, modern science, religion, and ethics into a heady stew of how human labor could quite literally conquer the stars. By today's standards you might call it a cyberpunk, transhumanist version of *Star Trek*, so nothing too modest.

Driven by these ideas and his admiration of the works of Jules Verne, Tsiolkovsky found employment as a mathematics teacher and began to build an enormous body of scientific work that would end up as more than four hundred papers on the physics and engineering of what would become space flight and space exploration. He was a fountain of prophetic ideas that included the fundamentals of rocket propulsion, the principles of steerable rocketry and rocket fuels, orbital space stations, the concept of airlocks and the essential elements of biological life support, and even the outrageous concept of space towers and elevators reaching from the ground up to orbit.

Presumably thanks to Fedorov, Tsiolkovsky also remained a staunch cosmist, writing in his 1926 *Plan for Space Exploration* that "men are weak now, and yet they transform the Earth's surface. In millions of years their might will increase to the extent that they will change the surface of the Earth, its oceans, the atmosphere, and themselves. They will control the climate and the Solar System just as they control the Earth. They will travel beyond the limits of our planetary system; they will reach other Suns, and use their fresh energy instead of the energy of their dying luminary."

Despite this enormous output of material, it was only toward the end of Tsiolkovsky's life—from the 1920s to his death, in 1935—that his work gained much greater notice and acknowledgment, especially in the Western world. In among his many efforts, one idea stands out in particular because it sits at the very heart of the physics of rocket propulsion. Tsiolkovsky describes this as the mechanism of "reactive flying machines," and although the British mathematician William Moore had explored this concept much earlier, in 1810, it was Tsiolkovsky's independent discovery and refinement that finally attracted broader attention. In fact, almost all records of Moore's earlier efforts were lost, rather ironically, in the destruction wrought by German bombs and flying machines in World War II.[22]

The underlying principle of reactive flying machines is that, as Newton had phrased it centuries earlier, for every action there is an equal and opposite reaction. Push against something, and you will tend to move in the other direction. Throw things away from yourself, and you will be thrust back by the kick of doing so. The exact amount of movement you experience can be quantified by keeping track of the momentum of everything involved—that mathematical product of mass and velocity whose conservation Noether proved is a consequence of nature's symmetries. The more mass you throw and the faster you can throw it, the greater the force—or thrust—and the greater the acceleration you'll experience in the opposite direction.

But if you propel yourself by throwing things away, you will also reduce your total mass with every throw, adding a complication to the bookkeeping of conserved momentum and acceleration. Tsiolkovsky provided the mathematics to describe the consequences by treating a rocket like a point of variable mass (variable because it is shedding stuff by throwing or blasting it in the opposite direction) and by using momentum to calculate precisely what your final velocity should be after a certain mass of propellant pushes itself away at a certain velocity.

Before it starts up, a rocket consists of its own mass (including a payload, like you and your spacecraft) *plus* the mass of the propellant that will be pushed away for you to gain velocity. Over time, the overall rocket mass gets smaller by the mass of propellant you've used. But the frustrating thing is that a rocket has to expend propellant to accelerate the remainder of its propellant until it too can be pushed out the back. A simple thought experiment helps illustrate just what a problem this is. Imagine that the mass of propellant that has to be ejected to accelerate and increase the velocity of a one-ton spacecraft to an orbital speed is also one ton. But in order to carry that ton of propellant, you will need even more propellant to accelerate what starts out as *two* tons of spacecraft and propellant. It's as if a nightmarish Zeno's paradox was created especially for rocket engineers.

Tsiolkovsky's rocket equation tells us precisely what the consequences are. There is a diminishing gain to the final velocity that you can reach by using more propellant because that propellant has mass and has to be carried and accelerated before being used. The actual equation—and a helpful diagram if you don't like equations—is shown in Figure 2.3, expressed in terms of the final maximum velocity that is reached (V_{max}) and the velocity of material in the rocket exhaust (V_r) multiplied by the natural logarithm of the ratio between the initial mass ($M_{Initial}$) of everything (rocket, spacecraft,

$$V_{max} = V_r \ln \frac{M_{Initial}}{M_{Final}}$$

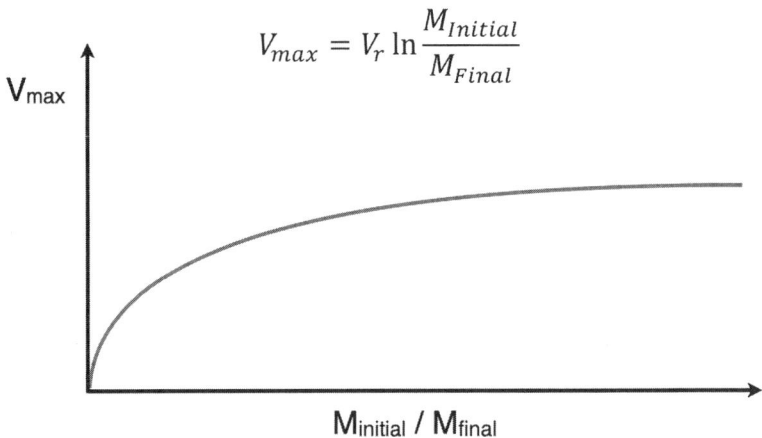

Figure 2.3. The behavior of Tsiolkovsky's equation is sketched in this illustration. As the mass ratio of the initial total rocket mass to the final payload grows, so does the final rocket velocity (V_{max}), but that velocity grows less and less easily with larger mass ratios.

and propellant) and the final mass (M_{Final}) after expending all the propellant.

To appreciate the profound implications of this otherwise innocent-looking equation and curve, it helps to consider two examples. First, suppose that you want to launch a rocket and reach a velocity that places you into a low orbit around the Earth. If you take off from somewhere near the equator and head toward the east to have help from the Earth's own spin (adding about 1,670 kilometers per hour, or 460 meters per second, at the equator), you need to gain a velocity of about 8 kilometers per second. If you and your spacecraft and the fabric of your rocket have a not untypical total mass of around 10,000 kilograms, then Tsiolkovsky's equation says that you would need about 74,000 kilograms of typical rocket propellant. That propellant will, in reality, often consist of two things: a fuel like kerosene, plus the liquid oxygen to burn it with in order to produce a roaring chemical-exhaust velocity of 4 kilometers per second. Bear in mind too that because the structure of the rocket has to house everything, the mass of that structure will grow with the amount of fuel and oxygen, *and* this calculation ignores the added effort needed to fight atmospheric friction.

That may sound like a lot, and it certainly is. Recalling how this chapter began, you now have an even more precise answer about why it's taken life so long to get into space. But this sort of launch is also clearly within the realm of possibility because we now do this kind of thing all the time. However, let's look at another, more extreme, scenario. Imagine that we instead want to launch something and send it to the nearest star to the Sun, a diminutive red dwarf called Proxima Centauri. Proxima is 4.246 light-years away, or about 40 trillion kilometers. That's quite a distance, so we're going to want to go as fast as possible to get there in a reasonable amount of time. However, we know it's going to be costly, so our compromise is that we're willing to travel for a hundred years. If we spent half that time accelerating and half that time decelerating in order to arrive and come to a halt precisely at Proxima, we'd need to reach a maximum midway velocity of about 26,000 kilometers per second (roughly 8.7 percent of the speed of light).

To keep our ambitions exceedingly modest, we'll further restrict ourselves to sending a single paper clip to Proxima Centauri. (I know that sounds strange, but you'll understand very shortly.) That paper clip has a mass of roughly 0.1 gram, so the fabric of our rocket can be made as light as possible. All that is left is to plug our requirements into Tsiolkovsky's formula to work out how much propellant a typical chemical rocket will need. The answer is 10^{2273} grams.

That is a big number: the digit 1 followed by 2,273 zeros. In fact, it represents 10^{2200} times all the mass contained in the observable universe. This is what it would take just to send a single measly paper clip to the nearest star in a hundred years' time. What does this mean? Well, it means that we can't do this. There is no way to use a chemical rocket to reach the nearest stars in any reasonable amount of time compared to a human lifetime. If we went a lot slower and took perhaps tens of thousands of years to get there, then maybe yes.[23] Or if we used a very different approach that doesn't involve carrying all of our propulsive mass along for the ride, we might begin to circumvent the problem. We'll discuss some of those ideas later on, but for less ambitious exploration, restricted to within our solar system, there are indeed some other tricks that can help a little bit.

For conventional chemical rockets, the most straightforward improvement is to make a vehicle out of detachable pieces: stages that hold the propellants. Each stage is then dropped or ejected as soon as it's empty, a technique employed by the majority of large launch systems today, perhaps most memorably by the giant Saturn V three-stage rockets of the Apollo program. Staging results in a significant reduction in the mass being propelled as time passes, at least by a factor of a few. The division of mass between the stages also allows for nuanced plans for when and where stages are separated. A first stage might be made bigger in order to power you high enough in the atmosphere so that when the next stage takes over, the surrounding air pressure is low enough to call for a different and more efficient rocket type and exhaust. For reusable rockets, staging lets you optimize things to help the components fly back safely. But staging also introduces many technical challenges and expenses, including having to build separate rockets for each stage, each entailing all the plumbing required to hold and pump fuel, control the system, and so on.

Some systems try to simplify this by using solid rocket boosters that are the direct descendants of the gunpowder-packed projectiles that people were using in the thirteenth century. A solid booster is generally a single cylinder of material with a hollow core that acts as a combustion chamber when the material is ignited and helps funnel the hot gases out through a rocket nozzle. The United States' space shuttle famously used a pair of solid rocket boosters to help lift off from the Earth, jettisoning these 45-meter-high tubes once they'd burned out so that they could be refilled and reused. Solid rocket fuel varies, but large rockets like those for the shuttle use a powdered metal like aluminum, together with a powdered oxidant that provides the oxygen to burn the metal. These powders are mixed with rubbery binders and then cast into the final cylindrical form—essentially forming a giant chunk of plastic explosive.

Like those home fireworks that you light with trepidation at arm's length, once a solid booster has ignited, there's not much control. By comparison, liquid-fueled rockets, pioneered in the early twentieth century by the likes of Robert Goddard, as I'll discuss, may be more complex but are also far more controllable. However, even rocket staging and the surreptitious use of solid or liquid rockets and boosters won't ever gain us enough to send that paper clip to Proxima Centauri within a hundred years.

The engineering of rockets is incredibly challenging, which is another factor in the whole natural history of space exploration. Liquid-chemical propulsion often uses inherently reactive substances that combine violently, converting chemical energy into undirected kinetic energy and momentum in a scorching gas. But then that momentum has to be coerced into creating as much forward thrust as possible. The chain of processes to make that happen takes place in a nearly continuous flow of whirring turbopumps, coolers, heaters, valves, and gimbals, and an awful lot of sensors and software. Large rockets will typically combine oxygen with a combustible component like kerosene, methane, or hydrogen. These compounds are pumped furiously into a chamber where they combust, expand, and are then funneled into the bell-like opening of a nozzle before whooshing off.

In order for a rocket to carry those ingredients efficiently, they need to be liquified. For oxygen and lighter gases like methane or hydrogen, that means pressurized propellant tanks and temperature control. The propellants are typically pumped into the tanks in a super-chilled, cryogenic state to reduce the pressures needed to contain them. This can involve a system that infuses deep-chilled liquid helium into the tanks to help maintain the low temperatures and to control pressure as the tanks empty. There is an added benefit of carrying all of these cold liquids, though, because they can be pumped around the rocket engine itself to help prevent it from melting as the propellants are combusted. Modern rocketry is in some ways all about high-stakes plumbing.

A big liquid-fueled rocket can pump a metric ton of material per second through pipes and chambers, urging that stuff on its way to explode outward and provide thrust without freezing, melting, or bursting anything. In fact, one key to making a good liquid-fueled rocket is to have the pressure of propulsive reactants in the combustion chamber be as high as possible. This means that you don't just need pumps to flow those reactants to the combustion chamber; you also need pumps capable of maintaining very high pressures. One way to do this is to siphon some of those reactants into a pre-burner that is in effect a separate internal rocket engine whose sole purpose is to drive the powerful turbopumps channeling fuel to the main rocket. It is an engine within an engine, and one of the Soviets' great engineering feats around the time of the Apollo missions (and part of the Soviets' failed N1 lunar rocket) was figuring out how to expel the pre-burner waste directly into the main rocket combustion chamber, adding as much as 25 percent lift power to the whole system.

Starting off in Earth's thick atmosphere and ending in the vacuum of space also calls for drastic adjustments to the way a rocket exhaust plume must be shaped and directed to maximize the thrust— the rate at which momentum changes. Rocket engineers also talk about a property called specific impulse that measures, in effect, the efficiency of a rocket. You can think of specific impulse as being similar to the measures of miles per gallon, or miles per kilowatt hour, that get applied to terrestrial automobiles. Specific impulse can be calculated as the change in momentum per unit mass of a rocket's propellant. The higher its value, the more thrust you get for a certain amount of propellant, and if you know what thrust you need and how long you need it to get a rocket to a desired velocity, you also know how much propellant is needed by just looking at the specific impulse. And specific impulse depends on the velocity of the exhaust as well as its composition, so you can gauge how much better things will be with a faster exhaust.

All of these steps, from calculations on a page (or in a piece of software) to the brutally abused alloys and chemicals in a rocket engine, end up generating a vector of ferocious forces targeting a tiny spot in order to send a machine onto a precise and stable arc of motion. Rocketry to get off a planetary surface is akin to asking an over-caffeinated monster to balance perfectly on one toe while hurling thousands of axes at a small, imaginary target all while on the back of a speeding truck that is plunging off a cliff. It looks majestic and easy only when all of those things are under firm control.

Using a continuous explosion for propulsion is also far from an obvious approach, despite its historical significance across centuries of gunpowder experiments. The violent reactions in this system transform chemical energy into kinetic energy in hot gases, but that kinetic energy results in matter moving in random directions—it does not naturally produce directional thrust. The superheated, high-energy gases of chemical propulsion have to be contained and channeled to turn chaos into a stream of momentum and thrust, with huge costs in efficiency. There is some gain if you can provide that kinetic—thermal—energy from something like a nuclear fission or fusion reaction to heat propellant, but it's still inefficient. A better option, in principle, is to purposefully accelerate all the microscopic parts of your exhaust in one direction from the get-go, something that can be accomplished using electrical fields to accelerate electrically charged ions of chemically inert elements like xenon into a forceful beam of thrust.

We have engines like this, called ion thrusters, that can work only in the vacuum of space, and they're incredibly efficient—reaching specific impulses ten times higher than their chemical-thruster counterparts (and were first brought to public attention by Tsiolkovsky in 1911). But without a big electrical power supply, they generate a tiny thrust that is practically useful only because it can be left switched on almost indefinitely in space. That works well for long-running robotic interplanetary missions like the Dawn mission, which operated from 2007 to 2018 and visited the large asteroid Vesta and the dwarf planet Ceres out beyond Mars, but it is pretty much useless for lifting off from a planetary surface. However, pair such engines with a bigger power supply, like a nuclear reactor, and in space you might be able to produce much higher thrusts and reach Delta-v's of an astonishing 100 kilometers per second. The efficiency gains also drastically reduce how much fuel is needed, and consequently how much extra mass you need to accelerate.

We'll revisit this and other propulsion options later, in thinking about the future. Some possibilities include keeping your "rocket" at home and stationary while it propels your spacecraft away from it—such as on a fountain of intense light from gigawatt lasers. Other possibilities might circumvent the entire notion of violent propulsion altogether by building structures that literally extend from a planetary surface all the way into space, or exploit the hair-raising dynamics of tumbling orbital "skyhooks" that might whisk you out of the atmosphere and into an orbit the way an angler hoists a fish.

But for now it's easy to feel what the astronaut Don Pettit has called "the tyranny of the rocket equation," the unholy unfairness of it all. That injustice doesn't quite end with the fundamentals of rocketry; it also exists for the wholly unsuited nature of most terrestrial life for anything but conditions like those on the Earth, and to some extent in the way that the history of rocketry has played out.

Although Tsiolkovsky's name eventually became synonymous with the theory of rockets, the story of practical rocket engineering has a long list of different names attached to it.[24] In the Soviet Union there was Sergei Korolev, born in what is now Ukraine. As a young engineer, Korolev survived time in Stalin's Gulag to become the Soviet Union's lead rocket designer and the driving force behind the pivotal launch of Sputnik 1 into Earth orbit in 1957. In the West we often talk about figures like Werner von Braun, who worked in the Nazi regime, using forced labor, to build deadly weapons like the liquid-fueled V2 rocket and was eventually "relocated" to the United States at the end of World War II to become a principal architect of America's space program, leading up to the Apollo missions to the Moon. Whatever else he was, von Braun had a bold, muscular vision of space exploration and expressed that in tomes like his *Das Mars-projekt* (*The Mars Project*). Written in the early 1950s, this outlined a huge mission to Mars consisting of 10 spacecraft and 70 astronauts staying for 1.5 years on the Martian surface and taking no fewer than 950 launches from the Earth to assemble an interplanetary fleet in space.

Elsewhere in the United States, a litany of rather more offbeat, homegrown engineers and rocketeers also made key contributions to the development of space exploration. There was Robert Goddard, a country boy born in 1882 in Worchester, Massachusetts, who worked his way up to a PhD in physics and independently discovered elements of Tsiolkovsky's rocket equation. By 1915, Goddard was building and testing rockets outside of his job as a physics instructor at Clark University.[25] He invented new ways to corral and channel the hot expanding gases from combusting gunpowder through carefully shaped nozzles in order to generate supersonic thrust—turning undirected momentum into highly directed momentum. Goddard had realized that it wasn't just the mass of material thrown backward from a rocket; it was the velocity that you could give to that exhaust that was so critical.

In 1926 he also launched the world's first liquid-fueled rocket, combining gasoline and liquid oxygen. It was a bizarre, spindly-looking contraption that barely reached a couple of hundred feet in altitude, but it was a breakthrough moment, right down to Goddard's improvised control panel for the launch, which consisted of telegraph switches mounted on a plank of wood—the great-grandmother of today's mission-control consoles. Famously, in 1920, after an article was published in which Goddard speculated on one day reaching the Moon, a *New York Times* editorial publicly and disparagingly dismissed his ideas about rockets working in space.[26] Forty-nine years later, the day after Apollo 11 launched for the Moon in 1969 using liquid-fueled rockets, the *Times* published "A Correction," in which it effectively apologized to Goddard (who had died in 1945). Goddard produced an astonishing 214 patents during his life and was often far ahead of his time. Notably, in 1916 he was already building and testing ion thrusters.

Others in the nascent field of space rocketry had even more exotic and unlikely stories. Take, for example, the truly bizarre tale of one Jack Parsons, who played a part in the very beginnings of what would become the Jet Propulsion Laboratory in California. Parsons has been called the "rocket-scientist-genius-occultist-playboy" of space exploration, which ought to whet your appetite.[27]

Marvel Whiteside Parsons was born in Los Angeles in 1914 to a wealthy but fractured family whose fortunes took a nosedive during the Great Depression. By the time he was nineteen, Parsons's adolescent fascination with science fiction, fantasy, and rocketry led to him taking work as an engineer in a gunpowder company, where he learned the intricacies of explosives and propulsion. By age twenty-nine, he had built his own solid-fuel rocket engine but had also branched out into a long-running family enthusiasm for New Age philosophy and occultism. Around this same time, he and a couple of fellow engineers taking classes at the California Institute of Technology finagled their way—with a professor's blessing—into forming a rocketry lab. This was before the term *aerospace engineering* really existed, and it marked the beginnings of what Caltech and the US government would spin off as the Jet Propulsion Laboratory in 1943.

Parsons and his group in the 1930s blew so many things up that they earned the nickname of the "Suicide Squad." But they built the first US rocket engine that ran for more than a minute, and they demonstrated rocket-assisted takeoffs for aircraft. They also invented a castable, composite solid-fuel rocket propellant that paved the way for more powerful launch systems that could reach space. Yet while all this intense, pioneering technical work was taking place, Parsons left his wife for her seventeen-year-old sister, habitually downed cocaine and methamphetamine, and seemed to have had a rather liberated sex life. He was also giving most of his money to an occult secret society called *Ordo Templi Orientis* that preached a head-spinning mix of ancient hermetic philosophy and magic, mixed in with dashes of Illuminati mysticism. He would even go on to engage in some funky dealings with L. Ron Hubbard, the founder of Scientology. Honestly, you just can't make these things up.

By the 1950s, the FBI had become extremely wary of Parsons's extracurricular activities and, despite his considerable contributions, banned him from any further work on rocketry. As a deterrence, however, this did little, and in a spectacular but awful finale he quite literally blew himself up in 1952 while working on explosives in his home laboratory.

Although the tale of Parsons is particularly colorful, it does capture an important aspect of the story of space exploration. Looking back at the human zeitgeist during the span of years from the late 1800s into the 1940s, it becomes a little clearer why modern rocketry and the journey to space began when it did. The horrors of new levels of global aggression and the growth of an industrial war complex were certainly huge factors in accelerating the process. But there was also a gathering wave of well-prepared scientists and off-the-wall thinkers who were convinced that a better future could be had in new cosmic pastures. Their single-minded drive pushed the principles and tools of space exploration to a new level, from the fictional insights of Jules Verne and H. G. Wells to the focused efforts of Tsiolkovsky, Goddard, Parsons, and a host of others. The concepts and mechanisms of rocketry and spaceflight were infusing themselves into the world as things that could actually exist.

Perhaps the revolution of Einstein's theories of relativity in the period from 1905 to 1915 also created a sense of possibility that the far reaches of the universe might be brought nearer. In Einstein's physics, space and time are nonabsolute, flexible properties of the world, and gravity is a product of the warping of the fabric of the cosmos by mass and energy. What had been immutable features of the universe became deformable and perhaps controllable. In this period of creativity, even the most well-established elements of classical Newtonian physics turned out be the subject of surprising new insights. The best example of this was an effect identified by the Transylvanian Saxon physicist Hermann Oberth in 1927: a discovery so simple yet so peculiarly nonintuitive that it was exceptionally well suited to the barrier-breaking Jazz Age of the time.[28]

The Oberth effect still causes even experienced scientists to scratch their heads. At its core are the concepts of kinetic energy and energy conservation, and their heritage through du Châtelet, Somerville, and Noether. Imagine—as in Figure 2.4—that a spacecraft starts at rest and then begins firing its rocket. As every second ticks by, the thrust of the rocket accelerates the spacecraft, adding another increment of velocity. The spacecraft's initial kinetic energy started at zero, and now its kinetic energy is growing every moment because that energy depends on the square of the rocket's velocity.

Let's look at this with numbers. To keep things simple, we'll say that the mass of the spacecraft and all of its propellants is 200 tons and that the rocket is very efficient, so as fuel is burned and expelled, the 200 tons figure doesn't change much during the first few minutes.

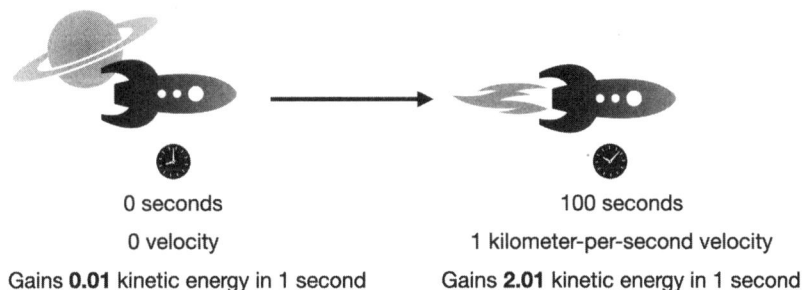

0 seconds	100 seconds
0 velocity	1 kilometer-per-second velocity
Gains **0.01** kinetic energy in 1 second	Gains **2.01** kinetic energy in 1 second

Figure 2.4. A simple thought experiment of a rocket gaining velocity helps illustrate the strange Oberth effect.

We'll also imagine that, at 1 second after launch, the velocity of the spacecraft has become 0.01 kilometers per second, so the kinetic energy is $\frac{1}{2} \times 200 \times 0.01 \times 0.01$, which equals 0.01 in some arbitrary unit.

At constant rocket thrust and constant acceleration, the spacecraft will then reach a velocity of 1 kilometer per second after 100 seconds of time has elapsed. At this point the kinetic energy is 100. Now the question is this: What will the kinetic energy be just 1 second after that? We can calculate this the same way and find that it will be 102.01. But this means that compared to when the spacecraft launched, when we gained a kinetic energy of 0.01 in 1 second of time, now we've gained a kinetic energy of 2.01 in 1 second of time—even though nothing has changed about the thrust of the rocket or the acceleration of the space-craft. In other words, 100 seconds into this flight we're now gaining kinetic energy 201 times more rapidly than at launch!

That is both the magic and the puzzle of the Oberth effect: faster-moving objects gain kinetic energy more easily.

This might seem like we're cheating physics; surely this violates rules against getting "something for nothing" and the idea of energy conservation itself. Well, it's not a cheat at all, but you do have to probe a little deeper to make sure. In the way that I constructed the previous example I've hidden some of the more complicated aspects that have to be thought through, but nothing that changes the fun-damental conclusion. If you're glazing over, skip the next paragraph.

One complication is that if you're riding along with the space-craft, you'll see the rocket exhaust moving away from you at the same velocity all the time. But if you were watching from elsewhere, like on the sidelines, as the spacecraft moves faster you'll see the exhaust leave the rocket with less and less velocity compared to your station-ary point of view. That's because the exhaust velocity is subtracted from the forward velocity of the spacecraft from your perspective. In fact, if the spacecraft eventually moves as fast as the exhaust does in the opposite direction, then the exhaust would appear to simply fall out of the back of the rocket and not be moving at all, like the stationary trail left by an acrobatic airplane writing smoke trails in the sky. Yet the total kinetic energy of the spacecraft plus the exhaust must be the same from all perspectives (another example of Noether's Theorem in action). So from your sideline point of view, the spacecraft will have gained more kinetic energy to make up for the apparently stationary exhaust—once again confirming that moving objects gain kinetic energy more easily. If all of that still sounds a little too good to be true, well, I didn't come up with the way the universe functions—I just work here.

Where Oberth's effect really, demonstrably matters is in rocketry and astrodynamics. Recall that if an object is in an orbit, then changing the kinetic energy will necessarily change the potential energy (to retain the balance of energies) and therefore change the size and shape of the orbit—as described using the vis-viva equation. But as we've seen, it costs to change kinetic energy in terms of Delta-v and rocketry's explosive efforts. The Oberth effect tells us that if there is a place in an orbit where you are moving fastest, this is the place you will want to activate your propulsion system to get the most bang for your buck in terms of kinetic-energy return.

In an elliptical orbit, that fastest-moving point occurs at periapsis, or perigee—the place closest to the object being orbited, whether a moon, a planet, or the Sun. Today's space missions tend to make full use of the Oberth effect by using their rockets at perigee to get the most efficiency in changing orbital trajectory. It is counterintuitive, but it is incredibly important. This works just as well for gaining velocity or for reducing velocity. For instance, when NASA's Juno mission needed to enter into orbit around Jupiter in 2016, it fired its rocket for 35 minutes to slow down just as it was whizzing by Jupiter the very closest and fastest, making the greatest change in its kinetic energy for the least effort.

Oberth's Jazz Age insight might seem like an incidental piece of niche knowledge, and in most respects it is for anyone who's not a rocket scientist, but it was also one of many signs that this was a turning point in history: a time when the mathematical rules of physics that our species had wrestled into shape could begin to be used to build real ships and devices to leave the world. The ideas of motion and gravity that had been percolating through the human psyche for thousands of years were on course to change the history of life on this planet.

— 3 —

THE HIGH PLAINS

All was stillness and desolation. Yet in passing over these scenes, without one bright object near, an ill-defined but strong sense of pleasure is vividly excited. One asked how many ages the plain had thus lasted, and how many more it was doomed thus to continue.

—Charles Darwin,
The Voyage of the Beagle

It's a little after 8 p.m. on July 20, 1969, and a fifteen-month-old toddler is being given a noisy, sudsy bath in the kitchen sink of his parents' modest apartment on the uneven top floor of a Georgian building on Great James Street in London, England.

This old capital is bedecked with many of our species' most distinctive cultural markings. The British Museum is just a few minutes' walk away, its dark iron railings cordoning off an imperious limestone facade. Inside is a sprawling entombment of the spoils and artifacts of human history and creativity that have been dubiously acquired in past years from across every continent. It's an astonishing collection that includes the strikingly stylized animals and godly pharaohs of ancient Egypt as well as some of the most exquisite prints and drawings from centuries of Western European art. All are a testament to the human urge to capture and communicate experiences of this world and others, whether real or imagined.

At a roughly equal distance in a different direction, yet more iron railings enclose the grand and grassy square of Lincoln's Inn Fields and its mix of trees: Indian bean, plane, ginkgo, holly, mountain ash, cherry, and laburnum. This park is another kind of museum, an open, living showcase of plant life whose sunlight-harvesting ancestors dotted the world as far back as the Middle Devonian period, some 380 million years ago. Here they've been cornered and coerced into this one place by our species and its simultaneous infatuation with and disregard for living things.

You can't see these places directly from the toddler's bath place, but they're part of his world and are often visited, often played in. Instead, out of the kitchen window, as he splashes in warm soapy water that's been siphoned from various sources along the estuarine River Thames, is a view of the newly built Post Office Tower—at this time the tallest building in the United Kingdom—with its blinking lights and cylindrical edifice making a rather lonely beacon of modernity shimmering in the warm summer air. This tower is, in its way, a museum to the future instead of the past.

My mother and father, because I was that squirming wet child in a kitchen sink, are completely distracted, paying no attention to any of these powerful markers of life and humanity on a July evening. Nothing could be further from their minds. In fact, all across the Earth parents are distracted and whole families are distracted. Entire nations are in temporary suspension, the normal flow of days and nights forgotten or ignored. At this very moment the collective breath of as many as a quarter of the human population is being tightly held.

At 8:17 p.m. London time, an angular capsule with thin aluminum walls sets itself down on the surface of another world some

383,000 kilometers away. That world represents yet another kind of museum, except this one is host to items from billions of years of alternate history. One of the exhibits is coal-black dust as fine as talcum powder, but as abrasive as sandpaper, that rests everywhere. That ancient coating billows into an angry cloud as it's suddenly affronted by an astonishingly noxious combusting mix of hydrazine, unsymmetrical dimethylhydrazine, and dinitrogen tetroxide that sprays from beneath the capsule to slow its final descent.

About 1.25 seconds later—the time it takes radio signals to traverse those 383,000 kilometers—my parents, and the rest of the planet, learn the incontrovertible fact that the "Eagle" lander of the Apollo 11 mission has touched down safely on another world.[1] In doing so, it has become the epicenter of humanity's base on the Moon. Perched on its skinny legs, the lunar lander sits in the southwestern corner of a large, flat plain that seventeenth-century astronomers had rather optimistically named the Sea of Tranquility, or Mare Tranquillitatis. But until the last few years, this great spread of desiccated iron-rich volcanic basalt, which spans nearly 900 kilometers, has never before, in its four billion years of history, seen a ship or vessel.

It isn't for another six hours that the powdery lunar dust stirs again. Before then, I am hauled from my bath, dried off, and put to bed. Years later, my parents tell me that they tried to explain what was going on, but my fifteen-month-old brain has no memory to offer of any of the excitement. Finally, at 2:56 a.m. London time, long after I am cosseted and asleep in my cozy, oxygen-rich atmosphere, astronaut Neil Armstrong drops to the lunar surface from the last rung of the Eagle's ladder.

Armstrong's warm, moist, multicellular body is encased in a suit made of twenty-two layers of material cocooning him and the three sets of undergarments he's wearing. Some of those layers include (from inside out) rubber-coated nylon, aluminized Mylar, the polyester fiber

Dacron, and a cloth of aluminized Kapton- and Teflon-coated fire-proofed silica fiber, all of which are encased in three outermost sheaths of fireproof and abrasion-proof layers.[2] It's a technological miracle, a flexible spacecraft for an individual human, showcasing just how far the ancient art of weaving has taken us.

Several of these shells are there to resist the forces that would otherwise make his pressurized suit pop open in the vacuum of space, with dire consequences for his tender body. Others are to keep him from being simultaneously roasted and frozen in the unfiltered lunar sunlight and shade. That thermal regulation is augmented with a web of tiny water pipes integrated into his undergarments for cooling and heating. Other layers are there to protect him from the rest of the solar system, shielding him against micrometeoroids that rain down onto the lunar surface. As tiny as these scraps of ancient orbiting protoplanetary rock are, they whiz through space at speeds where even a speck carries the heft of a bullet.

Deep inside this suit, Armstrong is as isolated from the environment around him as possible. He is a fleshy tube tightly sealed and packaged for preservation, using all the trickery and know-how that a civilization can muster. The only way that he can actually sense that his feet have sunk into the lunar soil is through its gentle resistance and a halt to his fall.

Nineteen minutes later, Buzz Aldrin, encased in his own spacesuit, joins Armstrong on the surface, and the two of them begin to carry out a tightly choreographed 140 minutes of tasks. They test scientific and engineering equipment, and gather a little more than 21 kilograms of lunar material to bring back to the Earth. They set up experiments for seismology and laser ranging and, of course, take pictures and plant a carefully tensioned US flag that unfurls artificially in the airless void.

It is an extraordinary moment, a culmination of a decade of intensely focused human effort involving hundreds of thousands of people and incorporating centuries of developments in physics and mathematics. This landing is a testament to our species' ingenuity and childlike bloody-mindedness in the face of an awesomely unforgiving cosmos. It is an instant that mixes poetry and inspiration with the parochialism and bluster of a species that evolution has spurted out as an experiment that is yet to be fully judged on its successes and failures.

Yet with Armstrong and Aldrin's descent to the lunar surface, the momentum of this colossal effort is already faltering. In 1969 the Cold War posturing that once helped set the entire Apollo enterprise in motion is starting to seem almost quaint and out of touch with the burning problems of Western society. The confluence of events (from the Bay of Pigs to a fear of losing out in space) that led John F. Kennedy to make his famous, rhetorically loaded speech at Rice University in 1962, where he stated that "we do this not because it is easy, but because it is hard"—thereby setting the entire $257 billion, 400,000-person Apollo program into motion—seems a distant echo, especially because Kennedy has been dead since 1963, Nikita Khrushchev of the Soviet Union was deposed in 1964, and the Vietnam war is finally dampening down.

Less than twenty-four hours after landing on the Moon, the astronauts launch back off the surface, rendezvous with Mike Collins, who is in orbit aboard the Apollo command module, and head back to Earth. It's about as touch and go as it can be, leaving behind a set of scientific experiments, together with the scorched landing structure of the Eagle, and a collection of rather uninspiring markers that includes eight receptacles of human bodily waste.

For a few years after Apollo 11, it still seemed possible that these excursions might evolve into something even more spectacular. To some extent they did, with Apollos 12, 14, 15, 16, and 17 being great successes. (Apollo 13, of course, failed to land on the Moon, but it didn't kill anyone.) By the time of the final mission of Apollo 17, in

1972, the crew spent more than three days on the lunar surface, where one of their sojourns outside the lander lasted for a record 7.5 hours and involved a 20-kilometer trip of the electric lunar rover—a triumph of technologically enabled exploration.[3] But this was the last human mission: after this, the visits of living, breathing, defecating people to the Moon stopped altogether. Politics, budgets, and attention had moved on to other things.

Forty-one years pass on the Earth.

I grow; I go to school. I get a degree, and I get a PhD. I'm as fully formed an adult as anyone gets to be. And now I'm sitting, more than a little nervously, face-to-face with Buzz Aldrin, the second human being to stand on the lunar surface on that day in 1969. Outside, it is a brilliantly sunny day in Battery Park, a district at the southern tip of Manhattan. This rocky spit of land is sandwiched between the continental mass of the United States to the west and the 150-kilometer stretch of Long Island to the east. The finger-like shape of Long Island reaches to a peculiar fork at its ocean-ward end, and its land is textured with rubbly spines and hills that break up its otherwise bland topography. These spines are evidence of the scraped and battered detritus from when the monstrous ice sheet of the Wisconsin Glaciation reached down across the latitudes to park as much as two miles of ice on top of the land and sea 2.6 million years back, retreating north only a scant 18,000 years ago.[4]

Although that deep history feels a little blurry and peripheral at this precise moment, it gradually comes into focus as a planetary coda to the seemingly improbable nature of meeting one of the few human beings who has been to another world. The meaning of many apparent coincidences is weighing heavily as I desperately try to focus. The clock that starts at my parents' kitchen sink in London in 1969 ticks off more segments, it seems, with every twist and turn of my life leading up to this point. All of this feels, entirely fallaciously, significant and critical.

Buzz's eyes are twinkling and robust, and he is disconcertingly, but unsurprisingly, not in the least perturbed by my inner turmoil. Our conversation has drifted in trajectory several times, and now we're discussing diving expeditions in Earth's oceans and the science of astrobiology—the search for life in the universe. We're enthusing about the prospects for exploring Mars and using its moon Phobos as a way station for astronauts who would control robots down on the planet. We are, in other words, having a happy, nerdy, scientific chat to which I'm struggling to stay connected.

An hour later, the meeting finished and polite goodbyes said, I'm standing rather dazed on the city's waterfront, staring at the Statue of Liberty off across the New York Harbor. I realize that for me, Aldrin is far more than just another famous person; he's also an embodiment of something extraordinary, something transcendent. He is one of the first human beings to become an interplanetary explorer. He is a bundle of a hundred billion self-aware neurons nursed into existence on the blue-ocean-world Earth and propelled outward to dance in the lunar dust, even if only for one day. What a day that was, though, and meeting Aldrin convinced me that although the Apollo program didn't carry on as it might have, its mere existence pushed exploration along a truly irreversible path.

I cannot possibly know what Aldrin's thoughts were during his time on the Moon, but I begin to convince myself that I was able to see in his micro-expressions and subtle turns of phrase a deep desire to once again set foot on those stunningly beautiful and terrifying high plains. I begin to imagine what it might feel like to gaze at the pristine glory of an alien place like that: a wilderness previously untouched by thinking beings for all of its four-billion-year history, where a person's desire to somehow register and confirm their existence in the cosmic scheme of things as a unique living entity can finally be realized. Whether Buzz was feeling any of this or it is just a product of my imagination, the whole notion creates an overwhelming sensation of awe in me.

You and I may have never set foot on another world, but with a little imagination and some contemplation I think that we can all experience a tiny amount of that frisson, that same thrill. That emotion continues to color our thinking about human space exploration—how could it not? There is something so primal about the idea of touching the universe, the reality beyond.

The Moon happens to provide life on Earth with a place that lends itself easily to the attention of those intense exploratory ambitions. Before Apollo 11's landing, there were seventy-one attempts to send probes to fly past the Moon, to orbit it, or to reach the lunar surface. Those attempts started in 1958, less than a year after Sputnik 1 was launched, although only twenty-four of those early Moon missions were fully successful. Since the Soviet probe Luna 3 (about which we'll learn more later) sent its grainy image of the lunar farside in 1959, we've learned that the Moon is a place of striking variation.[5] That farside of the Moon is almost entirely invisible from our vantage point, for the Moon's axial spin has the same period as its orbit around the Earth, keeping one face locked to our view except for a tiny sliver now and then as the lunar orientation shifts subtly. It's a battered wreck of a hemisphere that has been impacted and cratered by asteroids and comets over and over again. The familiar nearside is very different, with its vast and comparatively smooth plains of ancient frozen lava (basalt) like Mare Tranquilitatus, where Apollo 11 landed.

At the very outset of space exploration, the goal was simply to get close to or onto the lunar surface, at any cost. In fact, we had already smashed into the Moon before we had seen the farside, with the Soviet probe Luna 2 hitting the visible side of the Moon in 1959 and releasing a cloud of sodium gas that glowed enough to be seen by astronomers on Earth, helping confirm that an impact had occurred.

This was no small accomplishment: apart from the necessary rocketry and Delta-v to get there, just targeting the Moon well enough to collide with it at all was extremely challenging. Even if your spacecraft didn't suffer some teething failure or glitch, you didn't always have precise enough information about when and where you might actually intercept the lunar surface.

That general haphazardness led to one of the most peculiar legacies of this period: the inadvertent placement of two wooden spheres into orbit around the Sun that, as far as anyone knows, are still out there.[6] In 1961 the United States was trying to catch up with the Soviets' accomplishments, and it designed a sequence of spacecraft called Ranger 3, Ranger 4, and Ranger 5 that would take a payload to the Moon and send a set of special impactors down to the surface that carried instruments like seismometers. In order to give those sensitive devices a chance at surviving the crash, they were fitted into hollow cavities inside 65-centimeter-diameter balsa-wood spheres that were meant to absorb most of the force of the impact, bounce around, and come to a rest. The cavities were filled with fluid so that once the spheres settled, the instruments would float and right themselves. The fluid would then drain out and leave the devices properly oriented. It was ingenious engineering, but, sadly, things didn't go so well. Ranger 4 crashed onto the lunar farside and so couldn't be communicated with, Ranger 3 missed the Moon by 36,000 kilometers, and Ranger 5 had an electrical failure that resulted in it sailing past the Moon by a mere 700 kilometers. The by-product of these epic failures was that two of the balsa-wood spheres ended up on Sun-centered (heliocentric) orbits, where they likely still are to this day. It is a curious and wonderful thought that some unsuspecting trees, most likely from somewhere in Central America, were among the first organisms on Earth to find a way for a part of themselves to permanently break free of Earth's gravitational embrace, thanks to their usefulness to other organisms.

In the end, though, the relentless effort to target the Moon won out. Spacecraft like the five Lunar Orbiter missions flown by the United States were able to image and map 99 percent of the lunar surface by 1967. They used film cameras and onboard digitization to render resolutions as good as a few feet in places that were being scouted for potential human landings. The Soviet's Luna 9 mission performed the first soft landing on the Moon's surface and returned the first-ever image from the surface of another world in 1966. The Luna 9 data were also picked up by one of the radio telescopes at Jodrell Bank in the United Kingdom, where someone realized that the signal matched an internationally agreed-upon format for wiring images to newspapers. This allowed the signals to be converted to images that were quickly published across the world. This was probably a clever ploy by the Soviet mission designers to, in effect, "go viral" with the project's successes and circumvent their government's usual secrecy.

During this period of time, the United States was of course intent on beating the Soviets in the race to place humans on the Moon. The early Mercury program—a "Spam-in-a-can" approach to human spaceflight—gave way to the Gemini program, with its two-person orbital capsules serving as a test bed for the wholly new techniques that were needed for surviving and working in space. Those techniques included spacecraft orientation and maneuvering in orbit, as well as spacewalking and rendezvousing with other orbiters. At the same time, the Soviets carried out their own equally sophisticated—if not more advanced—program. In 1965 the Soviets completed the very first space walk, or extravehicular activity, where a cosmonaut floated outside of their spacecraft.[7] During this, in what must have been a very tense moment, the cosmonaut Alexei Leonov discovered that he had to release precious oxygen from his overinflated spacesuit just in order to squeeze back into the safety of the Voskhod 2 capsule.

Propelled by an extraordinary pace of development and expenditure across the 1960s, in December 1968 the Apollo 8 mission took humans around the Moon for the first time. A few months later, Apollo 9 undertook a rehearsal with the lunar command module and the lunar lander in Earth orbit. Then, just two months before Apollo 11, the Apollo 10 mission performed a complete dress rehearsal in lunar orbit, taking the lunar module down to within 50,000 feet of the Moon's surface.

Although the pace of large-scale lunar exploration slowed after Apollo 17's final human landing, in 1972, there have still been more than fifty robotic missions since then, either targeting the Moon or flying by to use its gravitational pull for orbital adjustments. The lunar surface has been mapped in visible light, infrared light, ultraviolet, X-rays, and gamma rays. Its rocky terrain has even been charted for its propensity for emitting neutrons as cosmic radiation plows into the dust and rock—revealing clues to the subsurface composition. There have been new robotic impactors and a few landers, and the rate of missions has picked up once again in the 2020s.

From all of this exploration we've learned that operating in lunar space is quite different than it is for local Terran space (to use the Latin word *terra* for our planet). For instance, the Earth's mass and atmosphere force us to place satellites at least 160 kilometers above the planetary surface, and many are far more distant, reaching out to more than 35,000 kilometers for precious geostationary slots with twenty-four-hour orbits. By comparison, the lack of atmosphere around the Moon allows for orbits skimming around at just 20 or 30 kilometers above the surface—at least in principle. The first real hint of a complication came in the 1970s, when two special experiments were carried out by Apollo 15 and 16. Before heading back to the Earth, each mission released a suitcase-size "subsatellite" into lunar orbit to study the radiation and magnetic field environment, and to refine our knowledge of the Moon's gravitational field.[8] These probes, called PFS-1 and PFS-2, were set in orbits that were about 100 kilometers above the lunar surface, but they subsequently experienced dramatic shifts in their trajectories. These changes were so much for PFS-2 that its orbit decayed and it crashed onto the surface just thirty-five days after it was released.

These seemingly erratic behaviors were not because of satellite dysfunction, though; they were a consequence of the Moon's deep past and its massively uneven and lumpy internal structure and gravitational field. At first glance, the Moon might look reasonably smooth on its surface. But like a hastily put together ball of clay and rock, it actually has enormous internal mass concentrations that are called, rather unimaginatively, "mascons"—that make its local gravitational field quite nonuniform.[9] For example, there are about five really large mascons beneath the lunar nearside, lurking under the smooth mare, or "seas," that you and I can spot by eye on a clear night. The largest of these mascons are dense enough that an astronaut walking across these regions would feel their weight changing by nearly 10 percent from center to edge.

Objects in near-lunar space are affected by the varying gravitational pull of these lumps, and consequently most close-in orbits wind up destabilizing over time, with some special exceptions. There are actually four orbital orientations, or inclinations, relative to the Moon's equator, where a satellite can sail on a steady course with the irregular pull of the mascons effectively canceled out. Orbital dynamicists call these "frozen orbits," and they vary from an orbit inclined from the equator at around 27 degrees to an inclination that places a satellite into a nearly polar orbit, circling north to south without instability. Consequently, low-altitude stable orbits around the Moon are scarce, presenting a problem for any future human aspirations of building satellite networks or other orbital structures. Lunar space turns out to be far trickier and more competitive than near-Earth space.

The lack of readily stable low orbits around the Moon can be a spaceflight headache, but it's also a direct window to a complicated and violent history that is intimately connected to the Earth's. The first real hints of that history actually predated space exploration and came from efforts to understand the nature of tides. Humans have long realized that the rise and fall of ocean tides, both day to day and across months, are associated with where the Moon and Sun happen to be in the sky. Although Muslim astronomers had charted these coincidences and proposed a direct, causal lunar influence since the twelfth century, it wasn't until the early 1600s in Western Europe that the astronomer Johannes Kepler formulated a proposal that the Moon was somehow forcing these changes—an idea that Galileo Galilei notably dismissed as rubbish. With Newton's work it became indisputable that gravity was the cause of Earth's oceanic tides as the planet spun through the gradients of the gravitational fields of the Moon and Sun. Furthermore, because gravity doesn't discriminate between oceans and rock, the solid parts of a world like the Earth must experience tidal forces as well—we just notice the oceanic tides the most. Finally, it was Laplace who produced a comprehensive, general theory of the time-varying, dynamic nature of gravitational tides in 1775, using the tools of differential calculus.

However, the physics and mathematical descriptions of tides are quite hairy when applied to complicated objects like planets that have distinct oceans and uneven rock. Consequently, it took time for scientists to unpack some of the implications for a system like the Earth and its Moon. One of the figures who played an important role in this was none other than Charles Darwin's fifth child, George Darwin.[10] Born in 1845, he had started out training as a lawyer but ended up as a professor of astronomy at the University of Cambridge until his death in 1912. One of his major interests was in the nature of tides, and he produced, among other things, a fearsomely mathematical volume in 1907 that collected his workings on tidal phenomena. One of the major puzzles that had emerged from tidal physics was that as tides move and distort the material of worlds like the Earth and the Moon, there must be losses of energy caused by friction, and there must be exchanges of momentum. In other words, a system like the Earth and the Moon has to evolve because of tidal effects.

Specifically, over time the Moon should be moving away from the Earth, to a wider orbit, and the Earth's day length should be increasing. Furthermore, the fact that the Moon always faces us in the same way—and its own day length is equal to its "year"—can be explained as a consequence of what is termed "tidal lock," where tidal forces have coerced the Moon into this arrangement. One clear implication of these behaviors is that the Moon must have once been much closer to the Earth.

This led George Darwin to propose that the Moon had formed from an expelled piece of a young, rapidly spinning Earth that eventually coagulated together to form our natural satellite. Today, we don't think that's the correct story, but Darwin's conviction that the Moon has a complicated history connected to Earth's turns out to be reasonably accurate. Direct measurements of the Moon's distance from us, using reflectors placed on the lunar surface that return timed laser pulses from the Earth, prove that it is indeed getting farther away by a few centimeters every year. We also know that, as well as being lumpy inside, the Moon's external shape is not spherical. It is an egg-like ellipsoid, and it appears to have frozen into this shape long ago, when the Moon orbited at half its present distance and was itself stretched more strongly by the tides of Earth's gravitational field.

The Apollo missions returned lunar samples revealing that the rocky contents of the Earth and the Moon also share almost indistinguishable isotopic ratios of elements like oxygen and titanium—a fingerprint of a very, very similar environment of formation some 4.5 billion years ago—so much so that the leading hypothesis for where the Moon came from is that it mostly came from us, echoing George Darwin's ideas. More specifically, the Moon and the Earth as we know them seem to have emerged from the collision of two proto-planetary bodies in the early solar system.[11] An object the size of Mars (labeled Theia) perhaps smashed into an earlier version of the Earth, a proto-Earth. This cataclysmic collision would have torn apart and mixed together these bodies. Yet in only a month or so, gravity would have regathered and condensed the pieces to form the Moon in orbit around a battered but still largely whole planet Earth.

We're still trying to figure out exactly how this collision might have played out, because even this scenario for blending two protoplanets struggles to explain how the Earth and Moon are so incredibly similar in their isotopic proportions. If the two original protoplanets came from even slightly different places in the solar system, their composition should be less alike. Some recent calculations also suggest that the Moon could have been produced in the very immediate aftermath of the collision and consequently be far more "of the Earth" than in other scenarios. Other recent research even suggests that there are massive structures that seismology detects deep in the Earth's interior that are remnants of that other world, Theia, quite literally parts of another planet inside this planet.

This history means that the Moon isn't just another solar system body. It is a chimera of sorts, a hybridization of two earlier planetary objects, and it is made of almost the same mix of elements that make our planet and form the molecules that make every living thing. The Moon may be separated from us by an ever-growing gap of space, but it is in all meaningful ways Earth's last and least explored wilderness, even though every human who has ever lived has known of its existence.

And as a wilderness it has some features that prompt a lot of superlatives. Although the farside bears most of the obvious scars from cosmic collisions, the entire lunar surface has been hit so many times by smaller bodies—from asteroids and comets to smaller meteorites and innumerable micrometeorites—that there is literally no part of it that has not been affected. In fact, the impacts and their craters layer on top of each other again and again in what is termed crater saturation.

Most erosion on the airless surface is from these ongoing impacts, grinding down rocks and pummeling material into the fine dust that lies everywhere. The Moon is literally an open-lid rock crusher. Counterintuitively, all of that weathering means that the Moon is also one of the darker and least reflective bodies in the solar system, with an average reflectivity, or albedo, of around 12 percent, compared to Earth's roughly 33 percent or Venus's brilliant 76 percent. The only reason that the Moon appears so bright in our skies is its proximity to us and the intensity of sunlight.

The highest and lowest points on the Moon, relative to its average radius, are associated with the mammoth South Pole–Aitken basin, a 2,500-kilometer impact crater mostly hidden from sight on the lunar farside. Part of the rim of this structure rises up in a mountain chain nearly 11 kilometers above the average lunar surface (and peeks into view from Earth at the lunar southern edge). A further crater within a crater within this vast basin has a depth of over 9 kilometers below the average lunar surface. This colossal geographic feature is thought to have formed when a 200-kilometer-wide object plunged into the Moon in a relatively shallow dive some 4.2 billion years ago. The early solar system was not a place for anything with fragile nerves.

This extreme polar region also has intriguing properties for space exploration. Some of those southern-polar-rim mountains poke up high enough that they are nearly continuously illuminated by the Sun—never shadowed by the Moon's bulk as it spins through its thirty-Earth-day rotation period relative to the Sun. If you needed to generate power with solar panels, this would be an excellent location for them. Conversely, just as there are permanently lit mountaintops, there are permanently shadowed crater floors.[12] In these regions, which have now had their perimeters carefully mapped, temperatures have likely remained below –170° Celsius (–274° F) for billions of years, in contrast to everywhere else, where, during the lunar day, temperatures can hit as high as 120° Celsius (250° F).

These chilly shadows allow for the possibility of substances like frozen water to persist even if exposed to the vacuum of space. In fact, these zones are at such low temperatures that they might act as "cold traps" where any free-floating molecules of water will attach to the surface and never escape again—like fog freezing to the ground. But the nature of water on the Moon is a strange and complex business. For a long time, following analyses of samples brought back by the Apollo missions, lunar rock and dust was assumed to be incredibly dry, an observation that appeared to make sense given the heat of the lunar days and the absence of any significant atmosphere. It seemed that water should just be baked out and broken up into its elemental parts to be lost to the vacuum of space.

But in the twenty-first century, technological advances in the measurement and analyses of those old Apollo samples, along with measurements from new space missions, changed that narrative. Fresh data revealed instead that water was bound into volcanic glasses and minerals in various locations and that there could be layers or chunks of water ice in places like the permanently shadowed craters. In some locations, like the Shackleton Crater (the spectacular crater inside the south pole's basin), data from orbiting sensors like those on the 2009 Lunar Reconnaissance Orbiter have suggested that a quarter of the floor of the crater might be coated in water ice.[13] The north polar region of the Moon also has permanently shadowed crater floors, and there are indications that a fifth of these areas could also hold forms of exposed water ice. Where all this frozen water comes from is another major puzzle. Some could originate from the deep past and the Moon's formation from those colliding protoplanets. Some might come from water-rich comets and asteroids that have pounded the surface. Other water could even form from hydrogen and oxygen that streams in from the solar wind—actual starstuff blowing away from the Sun's outer atmosphere—that combines in the lunar environment and winds up in those cold-trap regions.

Whatever its origin, though, if water exists in enough abundance anywhere on the Moon, it represents an extraordinary resource for space exploration. Humans need water to live and to grow food, but water can also be decomposed into oxygen and hydrogen by electrolysis, where—in essence—an electrical current is made to flow through two inert metal electrodes sitting in water. That would provide air to breathe and the fuel (hydrogen) and oxidizer (oxygen) to power rockets. Consequently, a place like the lunar south pole, with its alluring combination of perpetual light close to perpetual shadow, is under intense scrutiny by space agencies and commercial interests, and it is the prime target for exploration and possibly exploitation.

Although the Moon's surface should also harbor resources like rare-earth elements here and there, and perhaps helium-3 isotopes—implanted by the solar wind and thought in some quarters to be a key to easier nuclear fusion—the economics of refining and hauling stuff back to Earth are not great.[14] What is more important is the possibility of the Moon as a way station and a scientific laboratory: a place to iron out the kinks in longer-term exploration technologies and an easier launch point for exploring farther afield in the solar system—especially if water allows rockets to be refueled there.

The last decade has seen more and more countries trying their hand at lunar exploration. China has successfully flown several missions and has landed and deployed a rover at the south pole, as well as returning farside samples to the Earth. South Korea has deployed a lunar orbiter, and India has deployed two orbiters as well as a lander and a rover. Nonetheless, despite decades of experience, it is still extremely challenging to get a spacecraft to the Moon and for it to perform as we want. In 2019 an Israeli mission tried for a soft landing and crashed catastrophically. The United States has had a commercially built and operated lander hit major snags, losing its propellant after launch. Japan has flown small lunar "cubesats" built around an internationally standardized "Lego-brick" of readily packed

10-centimeter cubes that can be carried by other missions. One such spacecraft landed in 2024 with high precision on the lunar surface only to roll upside down—a posture captured in images taken from tiny spherical rovers codesigned by the Tomy toy company (maker of Transformers).

It may seem strange that exploring the Moon is still so difficult so long after the Apollo missions, but to a large extent this is because there are not many terrestrial endeavors that come even close to the complexity or raw physical challenge of exploration beyond Earth's orbit. On the Earth we've also had thousands of years, if not tens of thousands, to experiment and iterate on the technology needed for adventures like ocean faring. Space exploration is still relatively new, and for this reason we're not only still learning the basic skills; we're also still learning how the technological by-products of exploration directly affect the circumstances of our lives.

Pinpointing these influences takes some care, though, because the origins of some inventions or discoveries can get easily lost. For instance, over the years many things have been blithely and inaccurately attributed to the development of space technology by the United States—like the materials Kevlar and Teflon, or even now-common devices like the microwave. These products actually have entirely different origins; all were developed outside of the space program. But one of the most profound technological developments that really was directly attributable to a space program came from the decision to use compact microelectronics and programmable computers in the Apollo missions. Miniaturization of electronics was seen as the only way to pack the necessary computing power into a small enough space to be carried on the Apollo vehicles, at a time when computers could fill entire rooms. The need for that computing power came from the myriad systems controlling a human-capable spacecraft, but also from the complex orbital dynamics and situational navigation that a Moon mission calls for—and that continue to be enormously challenging today.

The necessary miniaturization was accomplished by accelerating the creation of "integrated circuits," which were the precursors of all of today's microelectronics. These tiny packages of sculpted semiconductor and conductor layers are built using photolithography, chemical deposition, and etching to form transistors, resistors, and capacitors. In the 1960s, companies like Fairchild in the United States could churn out thousands of these units, but the number that actually worked with 100 percent reliability was very low. The Apollo program couldn't tolerate that—these circuits had to be completely functional and robust. The Moon missions ushered in a revolution in the concept of electronic reliability and fault tolerance to meet the extreme demands of spaceflight.

But it wasn't just the hardware: the NASA scientists, together with contracted researchers at institutions such as the Massachusetts Institute of Technology, were also faced with an unprecedented software challenge. (Even the word *software* had been introduced only a few years earlier, in 1958.) The programs required to run the spacecraft—especially the lunar lander—had to monitor, control, and respond to a multitude of systems: navigational sensors and star trackers, spacecraft gyroscopes and distance-ranging devices, fuel pumps, electrical voltages across all the equipment, life-support systems, and more. These programs had to be capable of making real-time decisions, but they also had to inform the astronauts of those decisions and respond to specific human requests and overrides without messing up anything else that happened to be going on at the same time. While carrying out all of these tasks, the codes also had to coexist in a hardware "brain"—the Apollo Guidance Computer—that had 3.84 kilobytes of working memory (what we'd now call random-access memory, or RAM) and 69.12 kilobytes of fixed memory.[15] Those capacities are around a million times less than those of the average smartphone that so many of us carry around in our hands today. Furthermore, although fast by human standards, the Apollo computer also ran tens of millions of times more slowly than today's most ordinary phone or laptop computer.

Consequently, but not surprisingly, the code development for Apollo was a monumental task that took nearly as long as the development of the entire spacecraft ecosystem for the missions.[16] The engineers were walking in unexplored territory. The entire concept of human-machine interfaces as we now have it owes a lot to what the developers of the Apollo systems had to overcome. Perhaps most famously, the final system that was developed called for the astronauts to interrogate and control the lander's computer through an elegantly minimalist system that used two buttons, labeled "VERB" and "NOUN," and a numerical keypad. The astronauts would press VERB, then punch in a two-digit code for what they wanted to do (commands like "please perform," "terminate," or "display"), followed by NOUN and another two-digit code for what to apply that action or function to (categories like "range," "xyz planet," and so on). Once the astronauts had memorized the codes and practiced over and over in the spacecraft simulators, it was astonishingly efficient.

Today, this whole approach may seem primitive, but the engineers had no rule book or textbook to follow. They were in many ways creating from scratch the means for humans to engage with advanced computer technology. Along the way they invented software engineering and proved that machines, codes, and people could coexist. We see these same kinds of procedures reflected every time we fill out an online form or order, clicking or touching our way through menus and boxes. Once again, as with Darwin's influential theory of evolution, our compulsion to gain insights about the functioning of the world and the ideas surrounding those insights didn't just change the world; they also changed the way that the world could change.

It's not just human thinking that does this. The Moon is a strange and richly diverse place, and the more we look, the more we find new connections to our species' trajectory. This includes a wholly unexpected tale that takes us in a full circle between the rise of life on Earth and where we are now, with life perhaps poised to make its next move into the cosmos.

This story begins with the way that the solar wind plays across Earth's magnetosphere. The magnetosphere is an invisible region of space around the Earth spread across tens to hundreds of thousands of kilometers in a distorted teardrop shape. It's delineated by the ephemeral sheetlike structures of moving ions and electrons where the planet's magnetic field supports itself against the flowing electrons and ions of the solar wind. It's a little like the keel of a ship placed in a fast-flowing river. During about four days each month, the Moon passes through the tail of this invisible structure, and atoms that are stripped from Earth's outer atmosphere can stream through the magnetospheric sheets to land directly on the lunar nearside. Japan's Kaguya lunar orbiter, which operated from 2007 to 2009, encountered these sheets of particles many times. It appears that oxygen atoms that form in Earth's ozone layer and then diffuse upward to space are being carried to the Moon, where they end up implanted in the lunar soil.[17] We know this because those atoms have the isotopic fingerprint of oxygen from the Earth rather than oxygen from the Sun's wind.

This transfer has likely been happening for much of the past 2.4 billion years: across the time that Earth's atmosphere became filled with oxygen from photosynthetic organisms that learned how to harvest sunlight and across times when the Moon must have been even closer. Not only does this imply that the history of Earth's oxygen-producing life is painted into lunar soils on the Moon's nearside; it also suggests that there could be a way to examine how Earth's atmosphere is evolving, as new organisms—like humans—modify its composition and change its climate state.

Most intriguingly, though, this means that some of the water on the Moon forms out of these incoming atoms in the solar wind, which also means that it forms because life on Earth provided the oxygen to make it. If humans end up harvesting that water for the exploration of space, it will be a case of life providing its own means to move beyond the confines of a single planet and its natural satellite. Even if the contribution of Earth's oxygen to the amount of water on the Moon turns out to be modest, it is a story fit for a volume of cosmic parables in which a living planet thoughtfully stores its resources for a season yet to come.

Despite the extraordinary nature and potential of the Moon's wilderness, when Armstrong's deeply encased feet landed on its dusty surface in 1969, it was one of the sharpest reminders of the extreme differences between the environments of the solar system. Separated by just 400,000 kilometers and forged out of almost identical components, the Earth and the Moon are otherwise quite alien to each other. Furthermore, our wet, warm, nutrient-rich planet is incontrovertibly unique among its cohort, and its story is still being written day by day. That story is precisely what space exploration has been helping us understand since Sputnik 1 circled the globe in 1957, as we'll now see.

— 4 —

THE TERRANS

Well may we affirm that every part of the world is habitable!
Whether lakes of brine, or those subterranean ones hidden
beneath volcanic mountains—warm mineral springs—the
wide expanse and depths of the ocean—the upper regions of
the atmosphere, and even the surface of perpetual snow—
all support organic beings.

—Charles Darwin,
The Voyage of the Beagle

The launch of livings things away from the Earth began long
before the well-encased Apollo astronauts set foot on the lunar
surface. Even the very first suborbital craft, which were modified
V2 rockets propelled above 100 kilometers in the late 1940s, must
have carried an unwitting smear of microbial organisms.[1] The first
mammal to be rocketed above the Earth was a very unfortunate rhe-
sus macaque code-named Albert II (Albert I's fate is unclear), who
made it to the edge of what's considered space before plummeting
back to New Mexico with a failed parachute in 1949. Even before
Albert II, a sampling of fruit flies rode a similar V2-based rocket

in 1947 and were used to test the extent of radiation hazards in low-Earth space.[2]

In November 1957 the Soviet Union famously rocketed a small mongrel dog named Laika (who had been grabbed from the streets of Moscow) into orbit inside of what was only the second successful orbital device after Sputnik 1. Laika had what was perhaps an even worse time of it than Albert II. With no plan or way for her to reenter the Sputnik 2 capsule, Laika died after about two hours in orbit when the capsule's thermal control failed and its systems overheated. Since that time, the list of nonhuman species that have gone into space, and often into orbit, has grown and grown.[3] Chimpanzees have been lofted up to space, along with yet more dogs, cats, mice, rats, rabbits, plants, wasps, beetles, and even fungi. The first large organisms to pass around the Moon were a group of tortoises, along with some co-passenger flies and mealworms, all sent on the Soviets' Zond 5 in September 1968 (beating Apollo 8's human crew by a few months). The tortoises actually survived the trip and were recovered when their capsule landed back on Earth.

Bullfrogs, geckos, cockroaches, crickets, and spiders, as well as fish, sea urchins, shrimp, jellyfish, and squid, have all skated around the gravitational rails of orbit, together with all manner of other animals and plant life. When the Israeli spacecraft Beresheet crashed onto the lunar surface in 2019, it was carrying an experimental sample of a few thousand tardigrades, possibly the hardiest multicellular organisms known to science. The desiccated but potentially still viable forms of these tiny "water bears" may have been scattered onto the Moon's loose, dusty regolith.

By the 1960s, these experiments, often in cruel circumstances, had produced a general picture of what the space environment does to organisms that come from a wet, temperate, and massive terrestrial planet. The results are not all bad, but most of them are. In an orbit, weightlessness or near weightlessness profoundly affects organisms by triggering stress responses and a host of short- and long-term changes as biological systems search for a new equilibrium without an up or a down direction. Radiation is also a major source of concern. Particle radiation from the Sun, in the form of fast-moving protons and electrons, and so-called cosmic rays (from astrophysical sources deep in our galaxy that produce everything from gamma-ray radiation to ultrahigh-energy exotic subatomic particles) are largely prevented from reaching the Earth's surface by our atmosphere and are blocked to some extent by the close-up bulk of the planet. But in space it's always open season for radiation.

Putting humans into space brought these stresses and hazards into even starker relief. Although cosmonaut Yuri Gagarin's pioneering flight in 1961 lasted for only 100 minutes (including launch and reentry) and completed only a single orbit, his body showed stress responses right away. As people have stayed in space for longer and longer, the biological consequences have continued to come into focus. Today, more than 650 humans have reached space above an altitude of 100 kilometers, and of these, more than 600 have been in orbit, with two dozen of that cohort getting farther afield to the Moon. The cumulative amount of time that people have spent in space is around 180 years, with a cadre of some 50 individuals each with more than a year in accumulated total time. The absolute record holder for the longest single stay in space is, at this present time, the late cosmonaut Valeri Polyakov, who spent a contiguous 437 days onboard the Mir space station from 1994 to 1995.[4] Previously, he had spent 240 days in orbit aboard Mir in 1988, therefore experiencing the transition between the Soviet and post-Soviet space program from a rather perilous vantage point.

The scientific concept of "space stations," or semipermanent and permanent habitats in orbit, goes back to researchers in the early 1900s like Konstantin Tsiolkovsky and Herman Oberth, whom we met earlier. But in fiction they were beaten to it by the American author

Edward Everett Hale, who wrote a novella in 1869 titled *The Brick Moon*, in which a brick satellite 200 feet in diameter is launched into space and accidentally carries human passengers with it. Remarkably, Hale's fictional motivation for this artificial satellite was as an aid to navigation, also presaging the idea of global-positioning systems by nearly a century, as we'll see.

Hale writes about his protagonist observing the artificial world through a telescope and witnessing that the stranded humans have set up shop, even growing a hemlock forest under which to shelter. He also describes how "as I watched, I saw one of them leap from that surface. He passed wholly out of my field of vision, but in a minute, more or less, returned. Why not! Of course, the attraction of his world must be very small, while he retained the same power of muscle he had when he was here." In that one dramatic sentence Hale actually foresaw the bounding astronauts of the Apollo missions in the Moon's low gravity a hundred years in the future.[5]

To implement a real habitat in space is an enormous technical challenge, but we've learned a lot about how to do this. A simple example of this challenge is the seemingly very basic problem of how to get people on and off a space station, which turns out to be an extremely technical problem. This is a procedure that requires precision maneuvering capabilities and spacecraft control, together with an airtight docking between momentum-laden spacecraft as well as all kinds of fail-safe air-lock mechanisms. In any habitat there also has to be enough room for humans to function physically and mentally. Critically, there also has to be consistently breathable air and supplies of water, food, and power.

As with other aspects of the emergence of spaceflight in the twentieth century, a significant driver of space-habitat development was the possibility of having human operatives observing other nations' activities from space. Astronauts could be lookouts or camera operators and act as all kinds of intelligence agents from a few hundred miles up, where they were essentially impervious to inter-ference. For instance, in 1971 the Soviet Union launched Salyut 1 into a low 200-kilometer orbit.[6] This tubular spacecraft was about 15 meters long and 4 meters in diameter, and it had an assortment of pressurized and unpressurized compartments for people, machinery, and supplies. Though touted as a civilian space program, subsequent Salyut stations were further equipped for military work that moni-tored (one presumes) activities in the West. Some of the stations even included a gas-operated aircraft cannon for "self-defense" in the event that enemy spacecraft ever became a concern.

Altogether, there were six Salyut stations, based on the same design, that made it into orbit (another two failed to reach orbit), run-ning from 1971 to 1986. By the time of the last couple of iterations of the station, in the form of Salyut 6 and Salyut 7, these really were long-term habitats capable of supporting months of human habita-tion with two simultaneously docked Progress capsules for person-nel or cargo—that could also be flown automatically and uncrewed to and from the station. With these orbiters the Soviets gained an immense amount of practical knowledge about how to build and maintain a space station and how to keep spacefarers alive: from the mechanisms needed for air processing to recycling water. In 1986 this technological expertise was applied to produce a modular 130-metric-ton beast called the Mir space station that was assembled in space and operated for what was an unprecedented fifteen years from 1986 until 2001.

In the Western world, space stations had been a lesser focus—with human spaceflight efforts going to the Apollo missions and then the space shuttle. But, not to be outdone by the Salyut program, in 1973 the United States launched Skylab, a scientific-research habitat that used a modified version of the third stage of a Saturn V moon rocket. Though heavily publicized, it had serious deficiencies. There was no water recycling onboard, and during launch the station was quite severely damaged—leading to what was, at the time, a fairly audacious and successful repair mission by the first crew. Despite its recovery, Skylab was visited by only three crews, and by 1974 it wasn't in use. It reentered the Earth's atmosphere and burned up in 1979, scattering debris over Australia and the Indian Ocean.

Today, the only space stations in operation are the Tiangong station, built by China (following two earlier Tiangong stations operating from 2011 to 2019), and the International Space Station (ISS), built by a consortium of five space agencies and fifteen countries. The ISS has expanded since its inception in 1998 to become a monster of a structure that spans (with its large solar panels and trusses) roughly 100 by 73 meters—about the size of an American football field. Altogether, these stations mean that there are currently about 500 cubic meters of the universe beyond the surface and lower atmosphere of the Earth where humans can survive with some certainty.

One of the most important aspects of placing astronauts in these long-term orbital habitats is that we learn much more about what being away from the Earth does to us (as well as to all the other species we've dragged along). For starters, near-Earth orbit presents conditions of microgravity.[7] In microgravity you are mostly weightless, but there can be small gravitational tidal effects and nudges on a space station as it intercepts tenuous wisps of atmosphere. There can also be small changes in momentum when fellow astronauts move through the station and push off a wall, all of which can cause tiny accelerations. But for the most part, being in microgravity feels like endlessly falling. This can induce nausea, vomiting, headaches, and overall discomfort. It seems that this affects only about half of all spacefarers, though, and they soon overcome the problem.

What's much harder to deal with is a slew of other physiological effects from microgravity. Human cardiovascular and lymphatic systems evolved to handle fluids under Earth's surface gravitational acceleration, and they normally expend effort to stop everything from pooling in your feet. Put those systems into microgravity, and things get a bit haywire. Before their bodies adjust, astronauts can experience what they call "puffy head, bird legs" syndrome, where faces and heads appear inflated while legs thin down as fluids accumulate where they don't usually. Then there's increased arterial stiffness and plaque buildup, as well as longer-term myocardial fibrosis (scarring of heart tissue). The total volume of blood in a human in microgravity actually decreases over time, and hearts are prone to changes that reduce their capacity to pump.

Other organs get knocked around, too. Digestive systems can be impaired, gallbladders can become blocked, and space diarrhea is a real—and hugely undesirable—thing. (No jokes please, but yes, just imagine.) Livers can accumulate fat as if astronauts were high-intake alcohol drinkers. The lungs are challenged with mismatches that occur between their complex capillary blood flow and the diffusion of gases across cellular membranes. Human skin can suffer from rashes and psoriasis, and there are neurological changes in microgravity that include measurable loss of gray-matter volume—the volume of actual neuron cell bodies. Being in space can literally alter our minds. Then there is the rather-threatening-sounding spaceflight associated neuro-ocular syndrome (SANS). This is actually a number of distinct effects on the visual system that include microgravity affecting the eyeball and its contents, as well as various fluid changes in the head and subsequent impacts on the visual nervous system.

As if that weren't enough, immune systems have multiple reactions to the stresses of space. Latent viruses like those in the herpes family can even be reactivated—including the one responsible for chicken pox and shingles. Microbiomes (the tens of trillions of bacteria and archaea that use us, or any multicellular organism, as a habitat) undergo changes that we're only just beginning to understand. These changes seem to be induced by the stress caused by the conditions of microgravity but also by cosmic radiation.

Radiation in space really is an extremely tricky business. Even in low-Earth orbit, where the planet's magnetic field diverts some of the electrically charged particle radiation streaming through the solar system, radiation is immensely hazardous. As we'll see more of later, the Sun itself is a petulant creature, and solar activity in the form of flares and ejecta consisting of racing particles can be deadly. Onboard the ISS, for example, astronauts are drilled to know where the station provides more or less shielding against radiation and to shelter in the more heavily shielded parts when a solar storm is battering the Earth.

But just regular background radiation in an environment like the ISS can be quite insidious. The very-highest-energy (fastest-moving) subatomic particles that pierce our solar system have come from extremely powerful and violent generators across the cosmos—including exploding and colliding stars and colossal shock waves, and even the swirling matter around black holes. These particles are such microscopic juggernauts that they might zip right through a fleshy human without actually causing much damage, but—rather ironically—if they smash into atomic nuclei in the material of a spacecraft, they'll dump their energy into a spray of secondary radiation, like a fiercely hit ball in a game of pool hitting the rest of the racked balls. Those secondary particles can then devastate biological cells.

A less immediately worrisome though rather ominous phenomenon related to radiation has been reported by astronauts across the decades and especially by the Apollo astronauts, who went farther outside of Earth's magnetic protection than anyone else. This is the experience of perceiving noticeable flashes of light, even with eyes closed. The general consensus is that these flashes are induced by cosmic radiation that either generates photons inside the eyeball itself or triggers the neural impulses that create the sensation of light in the brain.

Careful monitoring of astronauts on the ISS indicates an average baseline radiation dose of about 300 millisieverts of ionizing radiation in a year-long stay. A millisievert is a measure of dosage that indicates the probability of cancer or other radiation-induced damage, so it is a complicated thing to calculate. Nonetheless, for comparison, in the United States the legal occupational dose limit is 50 millisieverts per year. Astronauts, cosmonauts, and Chinese taikonauts are not allowed to risk a lifetime dose of more than 600 to 1,000 millisieverts over their entire career—depending on what their home country has deemed acceptable.

Another dramatic effect of microgravity on humans is the loss of muscle mass and bone mass. Our skeletons begin to change almost right away, losing 1–2 percent of their overall mineral density for every month in space. Bizarrely, though, our skulls may actually get denser, whereas our lower spines and legs seem to suffer the most loss, at levels of 5–6 percent.[8] Astronauts can take anti-osteoporosis drugs to help mitigate some of this, but another effect is that as calcium leaches out of our bones, we run a greater risk of developing kidney stones. We don't actually know what the outcome is in the long term: whether skeletal issues stabilize or to what extent these issues really vary among individuals. Exercise that puts load on muscles and bones seems to help, but that's not easy in microgravity, and all manner of techniques have been tried and implemented. These include elastic tensioning systems to push while on a treadmill, or resistive cycling machines.

The vacuum of space also forces us into closed habitats that currently either are unhealthily clean and antiseptic or get overfilled with our own shedding microbial and cellular filth. Air is endlessly scrubbed and replenished to make sure that oxygen enters our bloodstream and carbon dioxide leaves it. Astronauts routinely take uppers, downers, sleeping pills, osteoporosis drugs, and anti-nausea medications. There is very little to living in space with present technology that is like our natural environment on the Earth's surface.

None of this seems very encouraging for space exploration, yet neither is it a coup de grâce. Humans, and other living things, can clearly exist for some time in environments outside of the Earth's. But to do so requires an enormous amount of attention and effort, as well as adaptations in behavior and outlook, along with medical and pharmacological intervention. We can indeed be in space, but in our present types of space stations and habitats, we'll need more than a toothbrush to survive it.

The technological advances that have placed humans and other living things into such alien but still cosmically local environments have another critically important side to them. That is in the advancement of the launching of machines and robotic avatars that in turn create new insights and forces that can fundamentally influence how life back on Earth operates.

In the fall of 1959, just two years after Sputnik 1 became the first artificial satellite of the Earth, a rocket launch from the coast of Florida propelled a modest 40-kilogram spinning top of a spacecraft called Explorer 7 into orbit.[9] A variety of devices were attached to this satellite to study the environment of near-Earth space. There was also a novel experiment that would, in retrospect, have profound implications for the past, present, and future of the entire human species.

That experiment was built out of five small hollow silver hemispherical sensors bulging out around Explorer 7's midriff. Like the sensitive bears in the tale of Goldilocks, these hemispheres were designed to respond to a range of extremes and just-right middles. But instead of the temperature of porridge, they were measuring the temperature of electromagnetic radiation. Two of the hemispheres were painted black to absorb sunlight and Earth's own infrared and reflected light. One was painted white to respond mostly to Earth's infrared light, and another was coated in gold to respond mostly to sunlight's shorter wavelengths. And one was tucked away from direct sunlight to sense only the light that the Earth reflected.

These hemispheres (technically termed "bolometers") would bathe in the radiation of the Sun and the Earth, and the different temperatures that the spheres registered would provide a way to assess the entire energy budget of our planet. Or, in the deceptively modest words of the original 1959 NASA documentation, "The experiment was designed to measure solar, reflected, and terrestrial radiation currents in order to obtain a clearer understanding of the driving forces behind the atmosphere."[10]

What the mission scientists captured from orbit was the first global assessment of Earth's energy balance: the near equality between the huge dynamic flow of energy coming into the planet from solar photons and the energy flowing out as the Earth reflects some of that sunlight and glows with its own heat. The researchers saw this balance changing with seasons and as cloud cover came and went, and they saw that the Earth was absorbing more energy than previously assumed.

These relatively simple measurements, made from 1959 to 1961, were unprecedented and helped set in motion a radical shift in how we thought about Earth's climate state and how we could study it. From orbit, we can, in effect, create a planetary thermometer—gauging the ebb and flow of energy of an entire world. A decade later, in the early 1970s, more technologically advanced descendants of this experiment would discover an imbalance in the Earth's energy budget: the flow out to space did not keep up with the flow in. This was a situation that could be attributed to ongoing changes in the very composition of the planet's atmosphere resulting from humanity's voracious appetite for combusting fossil hydrocarbons.

This wasn't the only early revelation to come from studying the Earth from space. In the spring of 1960, just months after Explorer 7 launched, a satellite called TIROS-1 was lofted into orbit.[11] This new spacecraft, whose full and rather awkward name was the Television Infrared Observation Satellite, was the first bona fide experimental weather satellite. It wasn't hugely sophisticated, and it basically carried a pair of wide-angle vacuum-tube TV cameras along with a narrow-angle version (all built by the Radio Corporation of America, or RCA) connected to a pair of video recorders that could store and then transmit footage to ground stations. With a sensitivity to infrared light, TIROS-1 was designed to capture images of cloud cover, and for the very first time, meteorologists were able to see the swirling shapes of Earth's huge storms and the changing patterns of clouds across the planet as the days and nights went by.[12]

In fact, although it can be hard to believe today, before TIROS-1 we simply didn't know just how organized cloud patterns were on a global scale or even what they really looked like. Never before had anyone seen the full breadth of the vast spiral structures of hurricanes and cyclones, the persistent features of equatorial weather systems, or even what happens to clouds at land-ocean boundaries. Suddenly, meteorologists had a way to accurately identify weather fronts from space and to make predictions about their future trajectories, utterly transforming how we forecast and prepare for weather.

These early projects heralded the start of a revolution in how we gather information about the Earth and what we understand of it. Prior to the space age, aerial photography and sensing systems such as radar or sonar had demonstrated how "remote" data could reveal a wealth of information about Earth's geosphere and biosphere, as well as about human activity. But placing remote sensing devices into orbit was a total game changer. In particular, there is one type of orbiting platform that has, along with its relatives and descendants, arguably done more than any other to alter the human trajectory itself.

In the United States, after a lengthy political and fiscal struggle between scientists and budget-conscious bureaucrats (as well as security-obsessed spy agencies), an innocuous-sounding mission called the Earth Resources Technology Satellite (ERTS) was placed into a near-polar orbit in July 1972, just five months before the final Apollo mission to the Moon. On board ERTS was a device called the multispectral scanner, designed by the engineer and scientist Virginia Tower

Norwood.[13] This scanner cleverly blended mechanically oscillating mirrors with electronic sensors to use the sweeping motion of the satellite as it orbited north to south over the Earth to build a continuous image of the planetary surface. This image was constructed through four carefully designed color filters corresponding to "spectral bands" ranging from visible light to infrared. At a resolution of about 80 meters on the ground, the map that was generated wasn't particularly detailed by today's standards, but the Earth's surface is 510 trillion square meters, corresponding to 80 billion ERTS "pixels." In 1972 this was an almost unprecedented flood of data.

By 1975, the ERTS had been renamed Landsat, and it would become the first in an astonishing line of Landsat missions that continues to the present day.[14] Even with the modest four colors of the original multispectral data, researchers were now able to peel apart the topography, geology, and contents of the Earth's surface in extraordinary detail. Different objects and different surface compositions reflect light differently. For example, healthy and unhealthy vegetation can be distinguished and mapped directly thanks to plants' photosynthetic pigments and a cellular propensity for rejecting infrared light. Using these data, scientists began to see Earth's living systems in an entirely new and exciting way.

The ebb and flow of global seasonal changes became clearly observable as geographic borders of foliage cover and growth pulsed in location from winter to summer zones. The impact of human activity was writ large in Landsat data, from urban regions coated in buildings to the geometric alignments of populations, industries, and the sinuous highways that cut across even the planet's remotest areas. For the first time we could also repeatedly gaze down on parts of the world where there may never have been human footsteps, from the vast deserts to the polar ice caps, as well as deep jungles and craggy mountain ranges. Data like these were, and continue to be, a source of revelation as we see the Earth's biosphere and surface technosphere vibrating with change in perfect time lapse from week to week and year to year.

Today, the ninth Landsat mission orbits our planet, as does the eighth. Each of these successive generations has incorporated better and better technology. The nearly 3-ton Landsat 9 platform can assemble 15-meter resolution images of the ground using its nine spectral band sensors, and it scans the entire Earth again and again over a period of just sixteen days.

Since 2008, Landsat's flood of data has been made freely available to anyone in the world. It's been estimated that national and local governments and other agencies have used these data to save billions of dollars per year in making critical land-use decisions and by being warned about emerging environmental problems as changes play out on the Earth's surface.[15] Landsat images help with agricultural risk assessments and monitoring water use by showing how the contents of rivers, lakes, and reservoirs fluctuate, and by pinpointing where lush green crops are being irrigated. Images are used for global-security monitoring, for tracking urban development, and for fire management and mapping forest growth and fragmentation. Vineyards use the data to manage crops and their own water use. Local authorities and other businesses apply the satellite information to issues like flood mitigation and to mapping out and analyzing agricultural commodities (with their different multispectral signatures), as well as coastal changes. The shifting zones of local climates and human activity are writ large across months, years, and decades. Even the habitat zones of migrating species like wildfowl can be tracked, along with the proliferation of fisheries and their impact on water properties and wildlife. Satellites like Landsat let us see the base feedstocks of Earth's ever-churning biosphere by tracking marine life and algal blooms in the most remote ocean locations—spotting sudden, mysterious changes or seasonal shifts. And we can track the ever-faster retreat of glaciers and ice shelves across the planet in near real time.

Of course, Landsat isn't the only game in town today. Scores of other remote sensing satellites have orbited the Earth since the early 1970s, run by different countries and international collaborations. Together they have profoundly changed how scientists and governments gather information. Different kinds of orbital instruments can provide detailed data on atmospheric composition and pollution or soil moisture and freshwater resources. They can also help pinpoint very politicized issues of state boundaries and drive the economic choices made in nations' development. More than eighty-five major remote sensing state-sponsored satellites currently orbit the Earth, and a growing population of private, commercial instruments are joining them to offer even speedier and higher-resolution localized data on a bespoke basis. There are companies like Planet Labs, which maintains an orbital swarm of some two hundred small satellites equipped with high-resolution multispectral cameras. A special subset of these devices can even distinguish features as small as 50 centimeters on the Earth's surface and may, for a price, be brought to bear on targets across the globe. Other subscribers can trawl the 50 petabytes of data that the company has accumulated over the past decade or so.

But anyone at all can access the Landsat archives through the Google Earth Engine, a cloud-based computing system that also provides access and software tools for a plethora of remote sensing data from this and other satellite systems.[16] From your screen you can probe Earth's weather and climate, using multispectral and radar data as well as a multitude of land-use measurements. Even the Earth's night is made visible, through data going back to 1992 that track the artificial illumination that our species floods the dark with. We've become rather inoculated against wonder in the ultra-connected, data-rich world that some 65 percent of all humans regularly access, but there is something astonishing about this planetary treasure trove. For the first time in human history and in the history of life on Earth, individual organisms can—if they choose to—base their thinking and decision making on an integrated knowledge of the entire planetary surface.

The oldest of all the remote sensing orbiters that still serve a purpose were launched in 1976: the incredibly long-lived Laser Geodynamics Satellites (LAGEOS), a pair of 60-centimeter-diameter spheres coated in highly reflective aluminum and 426 retro-reflectors. These superefficient mirrors were designed to allow pulsed laser light from the ground to bounce back easily from these satellites. That light signal allows us to monitor the spheres' precise locations in their 5,900-kilometer-high orbits and to use them as gravitational test masses for decoding the shape and distribution of Earth's mass, including the gradual shift of its continents. This technique is called geodesy, and it affects things like our understanding of how water flows across Earth's surface and how we navigate the planet's terrain. It is estimated that the first of these satellites, LAGEOS-1, will eventually reenter Earth's atmosphere in 8.4 million years as the ethereal drag of the outer reaches of air causes its orbit to decay. This bejeweled sphere also carries a plaque designed by the scientist Carl Sagan for whoever or whatever might receive this curious artifact in the distant future.[17]

We've invented many other forms of orbital remote sensing, too. For instance, one technique involves pairs of trailing satellites that refine the measurements of Earth's gravitational field to such an extent that we can make detailed maps revealing the anomalous lumps and bumps of different densities and materials far beneath the planetary surface. By monitoring the satellites' relative positional changes to the level of micrometers (millionths of a meter), these instruments can sense the movement of groundwater through its effect on gravitational accelerations. Elsewhere in orbit, radar satellites peer through clouds to track oil spills and sea ice. Lasers are used to probe the chemical constituents of the atmosphere or, together with radar, to act as a global altimeter with such precision that the surface bulges of Earth's oceans can be translated into maps of the rugged seafloor.

We can only imagine what a scientist like Charles Darwin would have felt if presented with all of these remarkable tools. His heartfelt exclamation about the habitability of all of the world's diverse regions in the passage at the start of this chapter is abundantly supported by what we can now see from space.

Of course, we Terrans don't just observe Earth's qualities from space; we also exploit space to communicate, to orient ourselves, and to keep an eye on one another. In fact, spying on one another's activities from space was, in many ways, the first truly large orbital investment for the United States and the Soviet Union. As far back as 1959, the United States had an enormous and enormously costly spy program up and running called Corona, which launched around 130 satellites between then and 1963. These "Keyhole" spacecraft mostly carried high-resolution 70 mm film cameras (some capable of imaging features as small as 2 meters across on the Earth) and capsules designed to be returned to Earth containing the exposed film in canisters. These reentry capsules could, in principle, be grabbed in midair by aircraft as they parachuted through the lower atmosphere, but if they were mislaid and fell in the ocean, they would float for a couple of days before a salt plug dissolved, causing the capsule to flood and sink to the ocean depths, taking its secrets with it. The Soviets had a similarly extensive program of reconnaissance, and both nations also built satellites to monitor radar and radio emissions, monitor nuclear tests, and serve as anti-satellite systems. Nowadays, even though the pace of launching spy satellites may have diminished, there are multiple nations running even more sophisticated and secretive orbital systems for watching one another.

We have also used satellites for real-time communications since the early 1960s. Satellites transmit data for phone calls and television signals across continents or around the globe. Often, these systems are placed in equatorial geostationary orbits, where a satellite's orbital period is precisely the same as the Earth's day length, and the spacecraft appears to stay almost motionless in the sky above a geographical region, allowing us to point our TV dishes in the right direction. Today, though, rapid evolution is taking place via the construction of vast satellite "swarms" or "constellations" at lower orbital altitudes to support high-speed internet and mobile-phone data by uplinking through orbital relays. The first serious attempts to do this for commercial use go back to the late 1990s and the start of the launch of the Iridium satellite constellation, which now consists of seventy-five active satellites. Other efforts are now rapidly exceeding these numbers. The Starlink satellite internet company (part of Elon Musk's many enterprises) already has more than 5,000 of its 300-kilogram spacecraft in orbit at an altitude of around 500 kilometers. Eventually, it might have 12,000 and may even expand this to a staggering 42,000 satellites. Others are following suit, with constellations ranging from dozens to thousands of individual spacecraft. All of these satellite swarms are designed to be endlessly replenished, because as their orbits decay, some spacecraft inevitably fail or reenter the atmosphere because of the forces of atmospheric drag from the tenuous outer wisps of Earth's air.

So many objects in low orbit around the Earth already cause a massive headache for anyone who values the view of the more distant cosmos. These satellites carry solar panels and reflective surfaces, and as they speed across our view, they can shine with reflected sunlight, polluting serious astronomical data as well as our individual experience of the night sky. Ironically, our expansion into space is also beginning to block our view of the rest of the universe.

Navigation is a close cousin to communication, and one of the most influential and widely recognized satellite functions is as a global-positioning system (GPS). The origins of this technology go back to Sputnik 1's launch in 1957, when observers found that they could track where the satellite was in its orbit from measuring the influence of the Doppler effect on its simple radio beeps. The Doppler effect is the change in measured frequency of a wave—including light waves—caused by the relative motion of the emitter of that wave and whatever is receiving it. If emitter and receiver are moving toward each other, the measured frequency increases, and if they're moving apart, the frequency decreases. In both cases, the size of the increase or decrease relates to the relative velocity of the emitter and receiver. This works in classical physics for sound waves and, with some Einsteinian tweaks, for electromagnetic waves as well.

The real stroke of genius was to recognize that instead of just locating where a satellite was in orbit, you could, in principle, flip this around to use the satellite locations to pinpoint where *you* were in space or on the surface of the Earth. To accomplish that, you also need a measure of separation from the satellites. Luckily, because the speed of light is a fixed thing—no matter what its source or receivers are doing—nature provides a ready yardstick. If you know how long a satellite signal has taken to reach you, then you can calculate the precise distance to the satellite. If you combine that distance with information about the satellite orbit, along with the same measurements from a bunch of other satellites, you can triangulate your own location. To make this possible, you just need purpose-built satellites in the right places and with the right equipment.

In the United States the first prototype satellite navigation system, called TRANSIT, consisted of five satellites launched in 1960 for the US Navy to help nuclear submarines pinpoint their locations in the thick of the Cold War. The subsequent story of the development of the modern GPS system (at least in the West, because the Soviet Union developed its own system and others have done the same, including the European Union and China) is lengthy and interesting in terms of social and political history.[18] But for understanding the implications for our interplanetary evolution and the possibility of Dispersal, it is the enabling science and the consequences that are most important.

Most modern consumer GPS devices, like your smartphone or car, primarily use radio signals from a constellation of roughly thirty active satellites run by the United States and twenty-four active European satellites. Each one-ton solar-powered spacecraft is placed in a roughly circular orbit at an altitude of some 20,000 kilometers above the Earth's surface. (See Figure 4.1.) The planes of these orbits are all tilted (inclined) away from one another and are longitudinally offset from one another so that at any given moment from anywhere on the surface of the planet there will be at least six US GPS satellites "visible" in the sky (barring mountains, valleys, and other nuisances).

The critical ingredients of each US GPS satellite include four rubidium atomic clocks, very stable radio transmitters, and some very elegant software and mathematics. The atomic clocks use the fact that

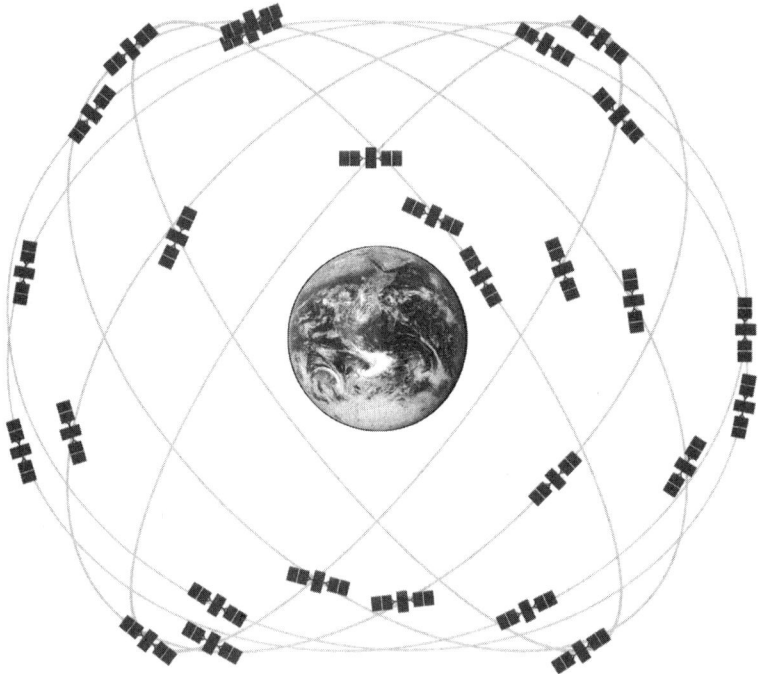

Figure 4.1. Rendering of the orbital configuration of twenty-four GPS satellites around the Earth (size of satellites greatly exaggerated).

electrons in atoms change their energy only by absorbing or emitting photons of very precise frequencies. By comparing the metronome-like beat of those frequencies to the ticktock of an electronic timekeeper (like an oscillating quartz crystal), it's possible to measure the absolute passage of time with exquisite accuracy and precision.

Each GPS satellite then transmits a continuous radio signal that includes a regularly updated time stamp, the satellite's orbital parameters, and code of unique but endlessly variable (so-called pseudorandom) ones and zeros that are generated from software that Terrans can also carry around in their GPS receivers, smartphones, or other devices. The pseudorandom code helps our small devices lock on to ("acquire") and distinguish among the relatively weak satellite signals from any number of GPS satellites at the same time. Once the satellite information and timing data are decoded, we can calculate the time it was actually sent and received according to the clock in our devices, and then calculate how long that signal took to reach us. Knowing what the immutable speed of light is, we can then figure out the distance the signal has traveled. With distances to at least three GPS satellites, a time from a fourth satellite, and a bunch of software trigonometry calculations, together with a host of extra corrections and details (like the Doppler effect and the effects of Einstein's relativity on the measurements, as well as atmospheric effects on the signals), the receiver can compute its position relative to the center of the Earth in three-dimensional space. Consumer devices like smartphones can do this with an accuracy of around 5 meters, but with extra equipment and trickery it's possible to shrink this to centimeters in real time or even millimeters over long periods. Consequently, there are some unexpected and extraordinary applications of GPS.

The GPS application most familiar to many of us is point-to-point navigation, which has played a role in the literal transformation of societies: tourism, delivery services, on-demand ride-share systems, aircraft and ship location, and managing entire transport systems.

Tracking devices that use GPS can even help retrieve your stolen goods or tell you when goods are being delivered. But GPS is also used to control essential machinery like farm equipment as it plows or harvests along exacting pathways, alerting farmers to hazards or helping map their crops. Elsewhere, ultraprecise augmented GPS is used in the construction of everything from houses to skyscrapers to road systems by providing centimeter-scale mapping and indoor-positioning information in minutes.

GPS location technology has been used in tags on wildlife, even on creatures like sharks, to help understand their movements and to know where human swimmers might be at risk of becoming a snack. For example, as I sit here on my sofa, I can tell you that a thirteen-foot-long adult white shark labeled "Bob" is currently enjoying life off the northeastern coast of Florida. I know this thanks to a GPS tracker and a website run by the global nonprofit OCEARCH, which keeps tabs on nearly five hundred sharks.[19] In other tracking efforts, GPS-enabled, solar-powered radio tags are placed on migrating bird species that allow them to be followed from the International Space Station, two hundred miles above the planet, beaming data back to earthbound scientists.

Long-term GPS data on fixed ground stations can achieve those millimeter-scale precisions to help monitor earthquake zones, tracking the gradual creep of fault systems like the San Andreas in California. And GPS satellite signals that reflect off the environment around you aren't just noise; they also represent data on the material contents of that same environment and can be used to measure properties like snow depth or vegetation cover, and even the moisture of soil.

Like the revolution of remote sensing from space, GPS has created a transformation not just in what we understand of the world but also in how we behave minute by minute. Living things have always used a multitude of tricks to locate themselves. The Earth's magnetic field is sensed by species as diverse as bacteria and birds in order to orient themselves and, through the fingerprint-like variations in magnetic-field line patterns, recognize terrain. Position-dependent chemical concentrations steer microbes and instruct the olfactory systems of many other species. The light of the Sun, the Moon, and even the stars is exploited by many organisms to provide directional information. And higher-functioning species encode consciously accessible memories of their environment in their brains to know where they are. But the emergence of a planet-spanning system that quickly, accurately, and precisely locates anything affords a species an unprecedented set of tools and opportunities. And because those tools and opportunities can involve the manipulation of environments—whether in building structures or carrying out agriculture or adapting around the movement of other species—they inevitably affect the rest of life on the Earth.

The measurement from space of Earth's inventory of life and materials helps track those changes. There are now orbiting devices that are descendants of those 1960s probes of energy and weather and that provide unique data on climate change and pollution.[20] Those data gauge properties like ocean salinity and temperature or track aerosol particles in the atmosphere from planetary processes or industrial pollutants. The information in these data paints a sobering picture of just how quickly we humans have pushed this rocky planet into conditions it has not experienced for millions of years. Also, the changes caused by our species are no longer confined to the surface realm of the Earth that we inhabit. Placing machines into orbit, still within the vestiges of Earth's atmosphere and magnetic fields, modifies that environment as well. Materials degrade, and chemicals evaporate into near-Earth space. Sunlight glints from surfaces that never existed in orbit before, and even biological structures now form a transient part of this strangely evolving environment.

Reaching space is clearly transformative for life on Earth, whether we look at the history of remote sensing of the planet and the satellite communications and GPS that have transformed the ebb and flow of organisms, or the ongoing expansion of living things into a narrow shell around the Earth and along a tiny thread of space to the Moon. Behind all of these things is a rolling, churning revolution in possibilities that is only gaining in momentum.

One gauge of the changing pace of those possibilities comes from looking at the broader history of space launches.[21] In 1960, just three years after Sputnik 1 reached orbit, the world launched 20 individual spacecraft. By 1965, that number had risen to 165 in that year alone. The rate of launches stayed fairly steady at more than 100 objects per year all the way until the year 2000. For the next decade it actually dropped slightly, although it still averaged more than 80 per year. Then in 2013 it surged to over 200, in 2017 it was over 400, in 2020 it broke the 1,000-per-year mark, and in 2022 it climbed to over 2,400 individual objects launched into space. In 2023 there were a record-breaking 211 successful rocket launches to Earth orbit or beyond, hoisting more than 2,600 new spacecraft into space. (See Figure 4.2.)

In other words, the overall rate at which objects are being launched into space appears to be undergoing exponential growth—where the amount of growth increases with each year. It's tempting to see the fluctuating rates of the previous decades as analogous to a pattern of establishing behavior and expansion that is all too familiar in biological populations. Of course, similarity does not confirm the same underlying mechanisms, but this may not be coincidental.

Much of the recent growth originates with billionaires (and other eager entrepreneurs) vying for orbital supremacy just as the imperial powers once did for the oceans. Well-backed companies such as SpaceX, Blue Origin, and Rocket Lab compete for a substantial cut of the increasingly large economic pie of space, as well as bragging rights and public exposure. Using a different outlook on risk compared to state-run space programs and traditional aerospace contractors, these

Annual number of objects launched into space

This includes satellites, probes, landers, crewed spacecrafts, and space station flight elements launched into
Earth orbit or beyond.

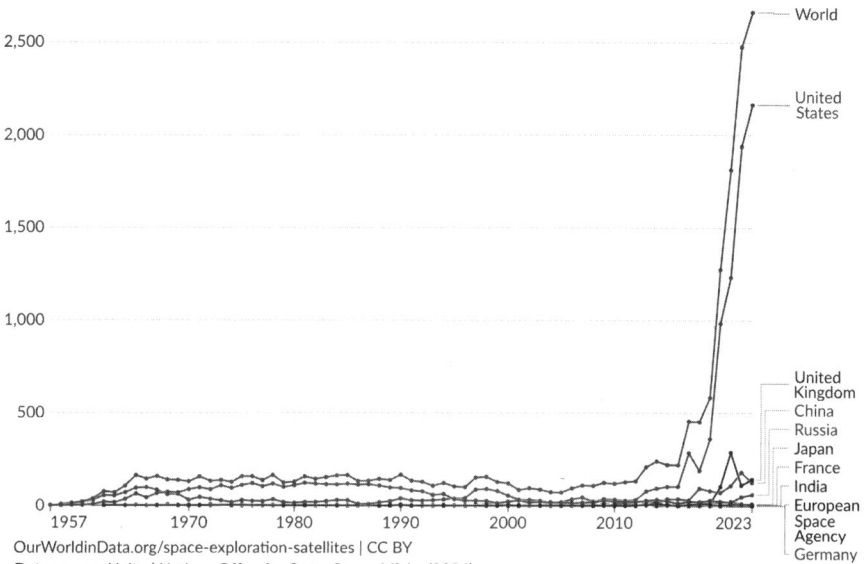

OurWorldinData.org/space-exploration-satellites | CC BY
Data source: United Nations Office for Outer Space Affairs (2024)
Note: Where they differ, launch attributions are based on the commissioning country, not the country
conducting the operations.

Figure 4.2. The estimated rates of spacecraft launches from the start of space exploration in 1957 until the present day.

efforts have brought a surge in technological advancements. Not least
of these has been the astonishing development of automated reusable
rockets and new human-rated spacecraft. For instance, SpaceX Falcon 9 boosters now routinely fall back to Earth under full control,
landing vertically and being processed for launching again. Consequently, some individual rockets have now flown more than twenty
times with only modest refurbishment.

In a grand scheme of life on Earth, these changes almost certainly
tell a story of a profound evolutionary moment, but it would be wrong
to not also scrutinize the more parochial, human aspects of the current growth in space activity. Not unlike those old imperial powers,
some of these companies are prone to rewriting the narrative of space,
maybe even gaslighting us into thinking that their ambitions are the

most exciting new thing for all of humanity. They leverage the inarguable romance and thrill of space exploration to gloss over the far less poetic financial and power gains to be had.

I prefer to not be cynical, but it's a sales pitch akin to the "imagining the future" story that technology companies have long used to justify selling us at least some products that we could probably live perfectly well without. Whether the private space industry is using this narrative cluelessly or deliberately, it is also a notable distraction from how wealth on Earth has become so astonishingly imbalanced. In the 1960s even the combined coffers of the richest half-dozen families in the world, from Getty to Rockefeller, could not have funded NASA's Apollo program. Today it would simply take a portion of the assets of the wealthiest person alive.

But not even billionaires can change the fundamental workings of nature, and the volume of space immediately around the Earth is not infinite. In 1978 the astrophysicist Donald Kessler pointed out that if the amount of material orbiting the planet exceeds a certain limit, it might result in a cascading failure.[22] Space debris—from defunct satellites as well as flakes of paint and other material—can collide, shatter, and produce more individual pieces that can in turn collide, exponentially increasing the likelihood of future collisions. This uncontrollable breakdown would produce so many orbiting pieces (especially in certain busy orbital zones) that it would be impossible to place new satellites without them being hit and rendered inoperable. Although the drag of Earth's outermost atmosphere can eventually deorbit untended objects, that natural cleaning process simply wouldn't keep up. Today, there's roughly 9,000 metric tons of debris in orbit, with some 500,000 pieces between 1 and 10 centimeters in size. Even though it hasn't happened yet, a collisional chain reaction event, known more generally as the Kessler Syndrome, is a very real concern.

Defunct satellites and other spacecraft pieces that reenter the Earth's atmosphere also cause terrestrial pollution. A 2023 study sampled aerosol particles of naturally occurring sulfuric acid in the stratosphere using a high-altitude aircraft and discovered that, in addition to metal elements and silicon from vaporizing meteorites and micrometeorites, there were metals from spacecraft in the air.[23] Specifically, there were twenty elements like aluminum, lithium, copper, and lead detected in concentration ratios that match the contents of burned-up satellites and rocket stages. As these pieces of technology disintegrate during reentry, they produce metal vapors that condense and combine with the other natural aerosols in the stratosphere.

Right now, the level of this pollution in the upper atmosphere is low, with around 10 percent of natural aerosols containing traces of these technological metals. But if the current rate of satellite launches and losses is extrapolated to the future, this contamination could rise to over 50 percent. We don't yet know what this pollution is doing, but as the metal vapors refreeze in the stratosphere, they may modify the formation of the aerosols—possibly making these particles smaller but more numerous. That could in turn alter the way that aerosols influence how energy flows into and out of the Earth, conceivably affecting the planet's climate and other systems. In 2024 a further study of the chemical effects of stratospheric aluminum oxide dust from burned-up satellites suggests that these particles could also catalyze reactions that deplete ozone, undoing all of our efforts to restore this vital shield of ultraviolet radiation. A Kessler event could cause a rapid and unprecedented surge of these pollutants.

These concerns have led researchers to start a few small projects that investigate the use of different materials for building satellites. This includes, rather surprisingly, the use of wood as a sustainable and less polluting option that simply combusts to carbon compounds when it reenters. A Finnish company has designed a small "cubesat" (the 10-centimeter-style modular system) whose main structural components are built with plywood, and Japanese researchers have opted for magnolia wood as a low-polluting option in their design.[24]

There's an intriguing side to these concerns, though. A Kessler-style disaster is likeliest in the lowest orbital ranges around the Earth. Here, satellites and debris are most closely packed and are whizzing around the planet the most quickly. Medium-Earth orbits, 2,000 to 35,000 kilometers above the planet, are unlikely to be affected anytime soon, and even the limited options for geosynchronous orbits (just beyond 35,000 kilometers) present far less collision risk. Although satellite tasks like internet connectivity, where signal travel times must be short, require orbits very close to the planet, the risk of the Kessler Syndrome seems to create a natural impetus to push outward—to expand the cloud of devices, habitats, and orbital laboratories around the Earth. That kind of evolution has parallels in surface environments, where organisms do this all the time to find resources and to reduce competition. Outrunning your own filth isn't always a failure; it's often the natural way of things. Maybe the Kessler Syndrome will actually encourage us to think bigger.

Although the thousands of satellites and tens of thousands of pieces of debris orbiting the Earth are still just tiny glinting motes—adding up to barely a millionth of a trillionth of the mass of the planet—from a cosmic perspective they're as revealing as any tiny living world you might find tucked into a grain of dirt or a drop of river water and examined under the lens of a microscope. These orbital specks tell a story of a planet where something interesting is going on, where billions of years of planetary agglomeration are undergoing a reversal of sorts.

Astronauts who have orbited among those motes have also discovered a shared experience in seeing the Earth from space. They call it the "overview effect": a profound sense of oneness with the biosphere that happens when you witness the entire crystalline majesty of the planet, an out-of-the-body transcendence that reveals and convinces just how precious, vulnerable, and interrelated we all are. With our robotic sensory augmentations, such as Landsat and weather satellites, we all now have ways to at least imagine this surreal omniscience. In a few seconds I can pull up Google Earth's satellite imagery to zoom into some of the most inaccessible parts of the planet and muse on what's going on there before checking on the happenings down the street in my own all-too-familiar neighborhood. The world is simultaneously much bigger yet much more accessible and *quantifiable* than it ever used to be.

Those quantifications, whether in the measurements of the energy balance of the planet or the measurements of land use and species migrations, have helped turn our planet into a place with far fewer mysteries than it once had. In the 1830s, as the *Beagle* sailed around the Earth, many parts of the world had remained unseen and untrodden on by any human for centuries, and perhaps even since modern humans emerged hundreds of thousands of years earlier. Today, although we may not have stepped on every rock or patch of soil, we can see pretty much everywhere, and we can measure what we see.

This colossal planetary inventory has irreversibly changed our conception of the world, so much so that all five hundred million square kilometers of the Earth's surface feel more graspable, more containable, and maybe even a little easier to live on than at any time in human history. Of course, we shouldn't fool ourselves that we really have that much control of a sextillion-ton mass of 4.5-billion-year-old stuff that is barreling around a star at 30 kilometers per second. But when a species reaches this point, there must be consequences in behavior and ambition, and history has already provided hints about the possibilities.

For example, there's a popular oversimplification of Darwin's intellectual journey that maps his experiences on the *Beagle* directly onto the emergence of the ideas of natural selection and the origin of species. In this telling, observations like the variations he recorded in the finches of the Galapagos Islands revealed the deep-time processes of selection to him. In reality, while Darwin's finches were very important, his studies of them yielded uncontrolled data that only hinted at a small portion of the extraordinary weave of fast and slow selective pressures and outcomes for these modest creatures (a story beautifully explored by the writer Jonathan Weiner in his 1994 book, *The Beak of the Finch*).[25]

In fact, to find supporting evidence for his ideas, Darwin went on to investigate domesticated species—reasoning that selective breeding caused by humans could work only if there was intrinsic, heritable variation within populations. For example, you can't produce a super-chicken if all chickens are identical to start with. Or, in Darwin's case, it wasn't chickens but "fancy pigeons" that provided critical data.[26] These domesticated pigeons have an extraordinary diversity in almost all traits: their beak sizes and shapes, whether their feet are scaly or feathered, and the way their tail feathers are structured. They even exhibit strong variation in behavioral traits like flight styles, where different breeds turn and swoop in unique ways.

Darwin hypothesized that all of these types came from a *single* wild ancestor, the humble rock pigeon, a fact that the breeders of the time absolutely refuted, fueling Darwin's urge to prove otherwise through his meticulous charting of pigeon traits and ancestry. But what is key here, and connected to what is happening with space exploration, is that in order to arrive at this hypothesis in the first place, Darwin was driven by a grander, universal picture. That picture was the nineteenth-century version of the overview effect, experienced by Darwin during the journey of the *Beagle* and during the explorations of his contemporaries, such as the naturalist Alfred Russell Wallace. That global vision of living systems and their ancestries led to a capacity to reexamine all the life right in front of us, including fancy pigeons. It turned these feathery curiosities into a major clue to the nature of living systems.

The twenty-first-century overview effect is having a similar impact. We live in the era of the "global" Earth, affecting everything from human economics to climate change to the health of species. But more than that, we're scrutinizing planetary function and its integration with living systems in new ways, inspired by the exploration of other worlds. Small curiosities, or seemingly mundane phenomena, are becoming important clues to the greater function of the world. Darwin sought to decode the origins and evolution of organisms and species, and he found that life's history also points to life's future. Today's overview effect and the tsunami of planetary data are extending that quest, but it is almost wholly about guessing the future.

The history of human thought and space exploration is a supremely important reference for scoping out possibilities yet to come, especially if we pose counterfactual questions about the past century of space exploration. For instance, we might have figured out rocketry earlier or followed through on the Apollo program and already pushed onward to Mars and beyond by the twenty-first century. But maybe there are excellent underlying reasons behind the turning points and collective decisions that we've made. Just as with any organisms that face a difficult new terrain, many supports had to be put in place first, even if we didn't realize the significance of what we were doing. Global weather monitoring and GPS, building planet-spanning communication systems, and being nudged along to invent robust microelectronics in order to get astronauts to the Moon all turned out to be critical for edging our way into the cosmos: pieces of a larger puzzle that each take on meaning after the fact. Given this, it's perhaps not too surprising that the very next pieces of the puzzle are about journeys that lead far, far away.

—— 5 ——

TO DISTANT SHORES

The night was cloudless; and while lying in our beds, we
enjoyed the sight (and it is a high enjoyment) of the multitude of stars which illumined the darkness of the forest.

—Charles Darwin,
The Voyage of the Beagle

Since the very first moments of spaceflight, scientists and engineers have been feverishly making plans for ever-more exploration
to ever-farther places. In 1958, barely months after the Soviets had
slipped into low orbit around the Earth with Sputnik 1, the United
States concocted an elaborate program to build four robotic probes
to quickly reach the Moon and Venus, with Mars also in its sights.
These early interplanetary missions were a far cry from today's highly
refined and exhaustively tested systems, but they had few budgetary
constraints or rules to follow because there was little by way of engineering legacy to rely on and immense political incentive for success.
The Soviets were likewise engaged, and just one year later, in 1959,
their Sputnik-derived, sphere-like Luna 1 probe flew by the Moon,
followed by Luna 2 impacting on its dusty surface and Luna 3 sending

back the very first images from the mysterious and unexpectedly rugged lunar farside. These glimpses of terrain unseen by any species for the preceding four billion years thrilled the world and were a technical triumph, but they were also entirely experimental and somewhat haphazard.

The Soviets were, by a hair's breadth, the first to encounter and solve some of the major problems of deep-space exploration—from the design and operation of space-worthy craft to interplanetary communications and navigation. Their innovations also took space exploration well beyond raw "get it there" functionality by tackling the challenges involved in carrying out scientific experiments and returning the results. For example, in order for Luna 3 to take and send pictures in 1959, the battery-powered spacecraft had a film camera that fed its shots to a tiny onboard development system.[1] This miniaturized darkroom doused the film in chemicals, dried off the negatives, and then scanned and digitized their images with a crude electronic television camera that transmitted the data back to Earth.

That media system, along with other instruments, was encapsulated in a sealed spacecraft body about a meter in girth. An atmosphere of pure nitrogen gas was circulated inside in order to regulate temperature and to prevent overheating, because in the vacuum of space there is no way to shed warmth except by radiating it away as infrared light. This is a far less efficient process than when a cooling breeze or ocean current carries away heat on the Earth, and heat management in spacecraft is, to this day, a critical but seldom celebrated engineering specialty. Topping off Luna 3's Cold War bragging rights, the specially temperature- and radiation-resistant photographic film used in the mission had actually been acquired from captured high-altitude US spy balloons drifting into Soviet airspace a few years earlier.[2] This was a fact that the United States failed to acknowledge, or was perhaps unaware of, while the world was agog at the Luna 3 accomplishments.

Beyond the Moon, the first full spacecraft encounter with another planet was a mission now often overlooked. In August 1962 the United States launched the Mariner 2 probe to scoot within 35,000 kilometers of Venus after a journey of just a few months. Mariner 2, as its name suggests, was already a second attempt. The rocket carrying Mariner 1 had been ignominiously aborted and blown up shortly after its launch went awry in July of that same year, a common occurrence in the early days of spaceflight. The Mariner family (of which there would eventually be ten members) consisted of machines built solely for operating in space and for interplanetary exploration, and when you see the design and complexity of these relatively modest early probes, you can begin to appreciate what's involved in setting out across the seas of our solar system.

Mariners 1 and 2 were particularly ungainly looking structures. Being in space requires few, if any, of the surfaces and symmetries of atmospheric travel.[3] These craft consisted of an open-tower framework that extended from a hexagonal base. This was the "bus" in spacecraft terms: the main body or structural element that holds everything together. Onto the bus were attached a pair of solar panels, radio antennae, maneuvering systems (a single rocket engine and ten pressurized nitrogen-gas thrusters), and some very basic scientific instruments to measure radiation, magnetic fields, and cosmic dust particles. There were also three spinning gyroscopes to help orient the craft, as well as basic sun and planet sensors to further help line the probe up correctly—the equivalents of the compasses and sextants of ships.

In fact, orienting and aligning spacecraft is a massive headache. After all, how do you decide which way is up or down or left or right, and how do you tell which direction you're actually moving in, and how fast, when you're all alone in interplanetary space? Even staying in a steady posture is hard when the slightest application of force is not resisted by atmosphere or gravity and can send things into a tumbling mess. To solve this, spacecraft are often deliberately spun on an axis to co-opt their own momentum and inertia into helping stabilize things. The Soviets and their counterparts in the United States pioneered the techniques to accomplish this and, perhaps even more importantly, expended the mental energy to figure out that these were problems that needed to be solved in the first place.

On probes like the early Mariners, the location of the Sun provided the easiest directional reference point, as well as being critical information for ensuring consistent power if a probe used solar panels to generate electricity. On Mariner 2 the spacecraft sun sensor had a simple but elegant design. It consisted of two small cadmium sulfide photodetectors that changed their electrical resistance depending on how much light fell on them. These detectors were carefully positioned across the span of the Mariner "bus" so that if the left and right solar panels were pointed properly at the Sun and each was being equally illuminated, the photodetectors would have equal electrical current flowing through them. If things got off-kilter, the shadowing of the detectors would cause unequal currents, and a corrective signal was generated that could be fed directly to the spacecraft thrusters to tilt things back into alignment. To catch more severe misalignments with the Sun, some of the later Mariners had six additional light detectors placed around their structures.

As well as being able to orient a spacecraft, you also have to know where it's going. There's an infamous story behind the Soviet Union's first attempt to send one of its probes to Venus, in 1961. That mission, Venera 1, suffered other mishaps that sent it barreling off into a silent and lost passage somewhere in the inner solar system. But it was perhaps doomed even before launch, because the Soviet estimates of where the planet Venus was expected to be in its orbit were off by 100,000 kilometers.[4] It's not something we tend to think of today, but in the early 1960s we really didn't know the positions of our moon or the planets with enough precision to place our spacecraft with anything more than a bit of careful guesswork and luck. It's one thing to watch these worlds in our night skies, but an uncertainty as small as a thousandth of a degree in our measurements of where a distant planet appears can translate into thousands of kilometers of misplacement in three-dimensional space.

In the 1960s the best option for properly locating Venus was to use radar measurements from the Earth, bouncing powerful microwave transmissions off the planet to gauge its position and motion, as if it were some massive submarine slinking through the cosmic night. Both the United States and the Soviet Union did this, but when the Soviet scientists announced their great new measurements of the Venusian position, their counterparts in the West quickly realized that there was an enormous error. It was so bad that the US scientists taunted the Soviets by asking out loud whether they had perhaps instead discovered a whole new planet. In this small leg of the space race it turned out that, much as in maritime exploration, having the best maps could make all the difference between success and failure.

The US scientists may have had a better idea of where Venus was, but at a mass of around 200 kilograms the first Mariners were already at the limit of what they could launch out of Earth's gravitational well. (See Figure 5.1.) Consequently, Mariner 2 had to forgo any kind of camera or imaging system that, in 1962, would have required considerable bulk. Instead, the spacecraft carried a set of rather dull-sounding but still scientifically essential instruments. These included microwave and infrared radiometers for measuring the brightness of radiation from Venus (not unlike the measurements being made of the Earth's energy flows at the time), as well as a magnetometer for sensing magnetic fields. There was an elaborate Geiger counter for monitoring subatomic particles that were streaming through space and a detector for measuring the impacts of tiny cosmic dust particles. The latter essentially consisted of a sensitive microphone attached to the spacecraft frame to listen for hits. Finally, there was a device for sensing the flow of the Sun's plasma—the mix of electrons and protons forever gushing away from our thermonuclear star as the solar wind.

Mariner 2 swept past Venus in December 1962, and these rather vanilla instruments produced stunning results. Scanning across the planet, from nightside to dayside, the radiometer data confirmed a nagging suspicion: Venus was stewing in an intense greenhouse atmosphere. Instead of a surface temperature similar to that of the Earth, the instruments suggested a fearsomely hot planetary surface, close to 500 degrees Celsius, simmering beneath a thick carbon dioxide atmosphere and opaque cloud layers that we now know are composed largely of sulfuric acid droplets.

This discovery would turn out to be prophetic for many places in the solar system that we had optimistically—or woefully unrealistically—imagined as perhaps being familiar and comfortable. The truth is that even if we've come to think of the named worlds as friendly features in the night sky, they are more typically brutal and unforgivingly alien environments.

These first forays, with Luna 3 passing across the lunar farside and Mariner 2 flying by Venus, were also early tests for the developing art of deep-space maneuvering. Luna 3, in particular, combined a "direct ascent" launch to the Moon—powering straight out from Earth and onto a lunar intercept—and what was, at the time, a daringly novel "gravity assist." In this instance, the assist used the Moon's gravitational pull as the spacecraft passed around the massive body to tweak Luna 3's trajectory, tilting its path and pulling it along so that it would end up orbiting *back* to the Earth and passing over our northern hemisphere. That return pass of the Earth allowed the Soviet radio communications station on Koshka Mountain in the Crimea to connect to the basic systems on the probe and receive seventeen out of the twenty-nine digitized pictures of the Moon.

Figure 5.1. An early
Mariner spacecraft, a
vehicle built to spend its
time entirely in space.

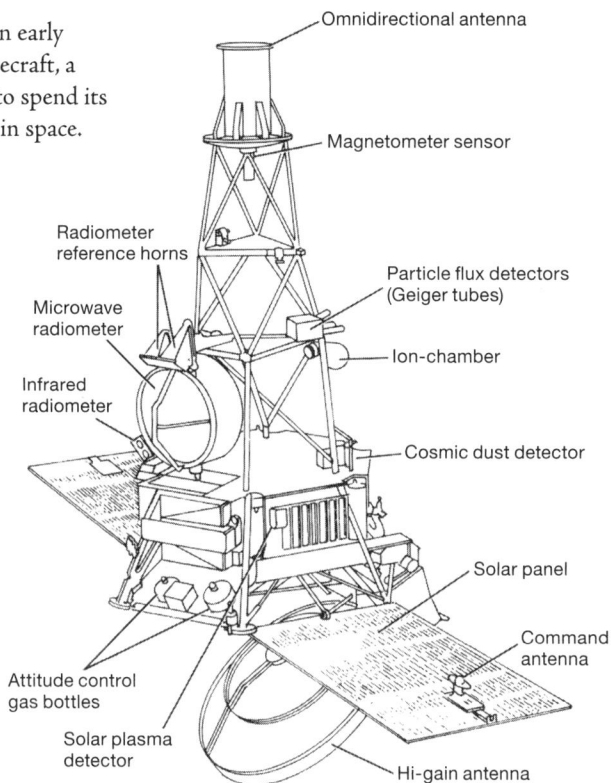

Omnidirectional antenna

Magnetometer sensor

Radiometer
reference horns

Particle flux detectors
(Geiger tubes)

Microwave
radiometer

Ion-chamber

Infrared
radiometer

Cosmic dust detector

Solar panel

Command
antenna

Attitude control
gas bottles

Solar plasma
detector

Hi-gain antenna

Mariner 1 spacecraft

This piece of orbital wizardry was an early demonstration that the earthbound rules of movement and exploration don't always apply in space. Although scholars like Newton, Laplace, du Châtelet, and Somerville had created and polished a set of mathematical tools for describing the orbits of planets and moons, it is a whole other thing to apply those tools to spacecraft and to understand the consequences for real-time exploration. The energetic expenses of rocketry and the currency of Delta-v are key pieces of the repertoire of space exploration. But these also give rise to a set of rules for motion and travel in the cosmos very different from our everyday experience of life on Earth. In a very real sense, how we—and all other life on this planet—usually move about and reach where we want to get to is a special case.

We usually experience one small part of a cosmic rule book of motion that applies only to circumstances in which gravity firmly pulls us onto a planet's surface.

Away from that comforting terrestrial attachment, we face a much more intricate and variable gravitational landscape. Things would be slightly simpler if you could escape to the depths of interstellar space, where it's possible to experience almost complete "free-floating," where gravity's tendrils are barely a whisper and nothing pulls you strongly in any particular direction. (However, this is somewhat of an illusion, for we are all embedded in the great Milky Way galaxy, which is itself barreling along in reaction to the gravity of other galaxies and even larger agglomerated structures.) But around the Sun, or any of the other objects in our solar system, there is never really any such calm.

We've seen that the most efficient thing to do if you don't want to fall into the Sun or crash onto an inconvenient planet is to ensure that you are moving in an orbit. If that orbit is long-lived, it will generally form a closed loop around objects like the Earth, the Moon, or the Sun. (Later, we will encounter classes of complicated, meandering, and nearly chaotic orbits that thread the invisible passages of forces between objects.) Being in a closed-loop orbit is the celestial equivalent of a holding pattern, so the question that arises next is how to get from where you are to where you want to go if that destination isn't on your immediate orbital path.

Even here on the Earth, we know that the way we get about in the world becomes very different when we're in the air or in an ocean. Ships that exploit the fluidity of the ocean and the atmosphere, like the *Beagle* with its sails and keel and rudder, get around in ways that most landlubbers don't find intuitive. The density and resistance of water enables a carefully rigged ship to push itself forward even if the wind isn't blowing in that direction. It does this by transferring forces from wind to hull to water in order to seek out the Newtonian magic of action leading to reaction. To tack into the wind this way, we exploit the material differences between the fluid-like flow of the atmosphere, the rigid inflexibility of wood or steel, and the hard-to-compress flowing density of water. That we and other creatures have learned to do this or have evolved into forms allowing versions of this kind of locomotion is actually quite extraordinary.

In an orbit the consequences of motive actions are, at first, even harder to intuit. A real-world example makes this very clear. Since the early 1970s, with the launch of Salyut 1, we've had space stations around the Earth, creating tiny habitable islands for spacefarers to maneuver toward. Even before this, in the 1960s, we'd performed orbital rendezvous, where one craft meets up with another. A particularly famous rendezvous was in 1965, when the American astronauts Jim McDivitt and Ed White were tasked with using their Gemini 4 spacecraft to reintercept the spent, still-orbiting upper stage of their launch rocket after it had hefted them into space and separated from their capsule.[5]

On the face of it, this sounded relatively simple: just gently use the Gemini thrusters to move in close to the upper stage that the astronauts could literally see out of their windows a few hundred feet away. But it proved infuriatingly difficult. Some of the problems were mechanical and physiological. For example, there were only two beacon lights glowing on the spent upper stage, making it difficult for the pilot, McDivitt, to ascertain its precise orientation in the stark light and dark of space. The astronauts also found that they had very poor depth perception without reference points, and they had no onboard radar to help sound out distances to the gently tumbling rocket stage. But the most difficult part of all was that when McDivitt, who was orbiting just behind and above the spent stage, fired the Gemini thrusters to move directly toward the target, he instead found himself sailing away from it—climbing higher and rapidly falling even farther behind.

You can just imagine what this felt like to someone like McDivitt, who was an extremely experienced test pilot and used to flying through Earth's atmosphere. In those terrestrial circumstances you essentially just pointed yourself at where you wanted to go and let the wings and jet engines behind you do the rest. In orbit this intuitive, reflexive process was all wrong, making for an incredible amount of frustration.

The problem was that in the rush of the United States' fledgling space program, McDivitt hadn't been fully instructed in the procedures needed to follow the rules of orbital mechanics. When he tried thrusting toward the target, the shape of his own orbit was changing. Adding Delta-v in the direction of motion propelled the Gemini 4 craft onto a more elliptical path farther from the Earth. The higher you are in an orbit, the slower your velocity and the longer your orbital period. For instance, in an orbit around the Earth at an altitude of 300 kilometers, it takes approximately 90 minutes to circle the Earth. If you raised your orbit to a height of 1,700 kilometers, you'd need roughly 120 minutes to go around. Consequently, instead of closing in on the target, McDivitt's actions were sending them ever farther away.

In fact, what he needed to do to bring Gemini 4 closer to the spent upper stage was to *slow down* in his direction of motion, dropping to a lower orbital path that would catch up with the target. If that still doesn't sound reasonable, well, welcome to orbital mechanics. The essential pattern of orbital-maneuvering steps can be traced all the way back to the rules of kinetic and potential energy that Émilie du Châtelet helped to crystallize in everyone's mind. In an orbit there must be a balance so that if, for instance, kinetic energy is altered by using a rocket's thrusters, the potential energy must change too, and that can happen only if the rocket moves either higher or lower around the planet or moon it is orbiting—precisely the source of McDivitt's frustrating day at work.

Today, a sequence of maneuvers for astronauts heading to the International Space Station might begin with a launch to a lower, and therefore faster, orbit than the space station. This is followed by adjustments to the orientation of the orbital plane so that it aligns with the station's. Then, when the capsule is about to catch up with the station above, it accelerates to change the shape of its orbit, moving it upward to within a distance where finer adjustments allow the craft to come together relatively easily. Although "easily" is perhaps a misnomer, because the same counterintuitive rules continue to apply, it is just that the adjustments needed in velocity and position come in smaller and smaller increments, and they benefit from decades of mission experience, as well as prompts from fast computers and computer vision.

If this all sounds like a lot of bother, you might reasonably ask why you couldn't, in principle, launch from the surface of the Earth and travel in an arc all the way right to a station and dock with it in one shot. You could indeed do this, except the maneuver would require a combination of exquisitely precise timing and very large and equally precise amounts of Delta-v applied by your rocket system, or else you'd basically just be a ballistic missile taking out the station and yourself. For these very reasons, this is not how it's usually done.

The travails of getting around in orbit around the Earth are one thing, but going farther afield creates many possibilities and many equally counterintuitive situations. For example, the elliptical Hohmann transfer orbit that is often used to travel to Mars may be one of the simpler examples of an interplanetary route plan, but it comes with its share of complications. Because the outermost reach, or aphelion, of a Hohmann orbit around the Sun is designed to precisely match the distance of Mars, the transfer must be timed to bring a craft to that point just as Mars is scooting along in its own orbit around the Sun. Mars takes 687 days to circle the Sun, and the Earth takes 365, so there is a rhythm to variations in the length of that transfer path that is described by the *synodic period*—a stretch of time between the planetary conjunctions when these two worlds are closest together. Consequently, there are only certain years and calendar dates where we can launch from the Earth and make the most efficient rendezvous with Mars. Roughly every 780 Earth days we can get to Mars the most quickly, taking about six months. Miss that window, and the travel distance and time can be significantly more—unless you expend a huge, and for now impractical, amount of Delta-v.

Another tricky set of choices faces us if we send people to Mars and expect to also bring them back again using a transfer orbit. It turns out that a short stay on Mars of about a month might force a much longer return journey—actually having to loop around the Sun before getting back to the Earth—while a long stay on Mars allows the astronauts to wait for a quicker Hohmann transfer back home. In truth, it's all like having to plan your interplanetary life around the most rigid and inconvenient bus timetable imaginable.

But the intricacies of getting to and from Mars are, by celestial standards, quite simple. In the complicated gravitational landscape of our solar system, amid the constant motion of all objects, there are some very strange places to go to and ways to situate yourself. One of the most famous examples lies in the so-called Lagrange points. Discovered mathematically in several stages by Leonhard Euler in the 1750s and by Joseph-Louis Lagrange in the 1770s, these points are five special locations in the plane of a system of two massive— but unequal—orbiting objects, such as the Sun and the Earth or the Earth and the Moon.[6]

Obligingly, like many other orbital phenomena, Lagrange points are fairly counterintuitive. They exist because of the rotation of the system—and its centrifugal forces—and the inverse-square-law drop-off of gravitational forces with increasing distance, two phenomena that we don't really have to deal with in our everyday lives.

Lagrange points are places of equilibrium where, in effect, the forces of gravity and the orbital rotation of a system of massive bodies conspire to create patches of calm where you can sit and stay with minimal effort and always be in exactly the same relative location to the massive objects. Some Lagrange points are like curved saddles in terms of the potential for small things like spacecraft or asteroids to drift away (and are therefore calm but unstable places to sit without help); others are more akin to smooth hillocks in an otherwise endlessly sloping and slippery landscape. On these imaginary hillocks, which are also unstable equilibria, objects that "roll" away find themselves subject to Coriolis forces (because everything is in orbital motion) that actually keep them perpetually moving around the hillocky Lagrange point. That can all seem a bit confusing, and the math is not particularly simple, but the key is to imagine these places as spots of still water in an otherwise choppy ocean.

The first Lagrange point, or L1, is a saddle point that lies directly between the two massive objects in a system and moves in perfect synchronicity with their motion, staying precisely in the same relative spot. It is, in effect, the place where the opposing gravitational pulls and the acceleration of orbital motion of the two objects balance out. For example, for the Earth and the Moon, their L1 point lies at about 85 percent of the distance between them, measured from the Earth.

The second Lagrange point, or L2, is another saddle that lies along the same line but on the far side of, in this example, the Moon from the Earth, and it is also fixed in its location relative to the massive bodies. The third is again along this imaginary line, but now 180 degrees away on the far side of the most massive body, the Earth. And the fourth and fifth—the L4 and L5 points, which are hillocks—sit at fixed distances ahead and behind in the orbital path of the less massive body, the Moon, at a sixth of the way around the lunar orbit in either direction. Figure 5.2 helps illustrate where these points are.

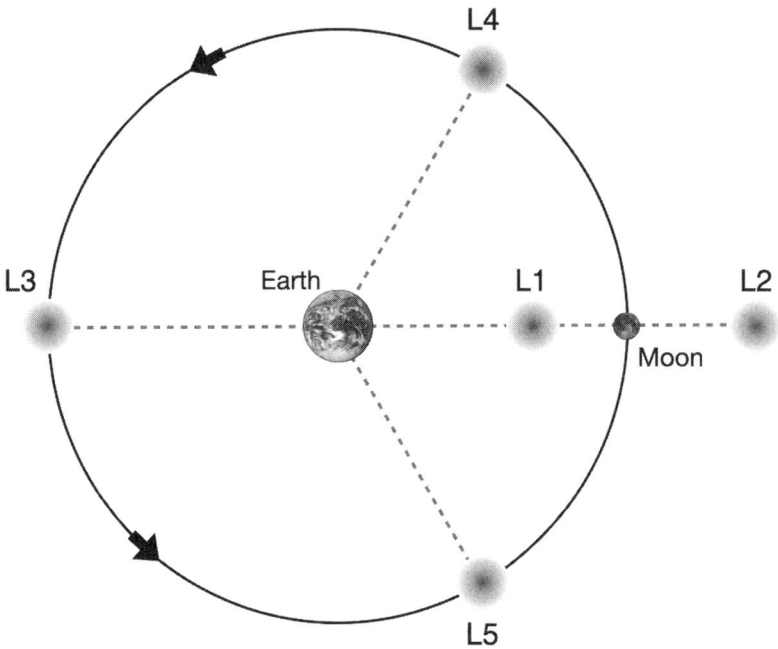

Figure 5.2. Illustration of the Lagrange points in the Earth-Moon system. The Moon and Earth are constantly moving in their orbits, but the locations of the L1, L2, L3, L4, and L5 points relative to the Moon and Earth all remain the same at any given moment.

Our minds might struggle with the concept of Lagrange points, but nature is very good at finding these places, especially L4 and L5 points, where small objects can roll around indefinitely. For that reason, they are places where material can accumulate over time.

For the Sun-Earth L4 and L5 points, that gathered material is interplanetary dust and a couple of persistent asteroids. Even in the more modest Earth-Moon L4 and L5, there are notable concentrations of dust, the small particles effectively trapped for a while. But the most dramatic example of the gathering power of these two points is in the Sun-Jupiter configuration, where thousands of asteroids fill these locations as leftovers from the formation of the solar system. The asteroids that lead Jupiter in its orbit are said to belong to the Greek camp, and those that trail Jupiter belong to the Trojan camp, in homage to Homer's *Iliad* (although both sets of these asteroids tend to be collectively called simply "Trojans," perhaps a small compensation for the losing side).

Right now, NASA's Lucy mission is en route to visit a few of these Trojan asteroids, meeting up with them in the 2030s. The name Lucy was chosen as a reference to the moniker given to the fossilized skeletal remains of a female hominin of the Australopithecus species who lived 3.2 million years ago in what is now Ethiopia. The existence of that fossilized terrestrial Lucy has helped us reconstruct a part of the history of our origins, just as visiting the Trojans may help us reconstruct a part of our planetary origins.

Clever little hominins that we've become in the intervening millennia, we've also learned to use Lagrange points more directly for ourselves. Over the past decades some twenty spacecraft have parked themselves at, or made use of, Sun-Earth and Earth-Moon Lagrange points. That includes the behemoth space telescope JWST, currently keeping station in a 200,000-by-800,000-kilometer loop called a halo orbit, around the Sun-Earth L2 point.[7] This is a place where JWST needs very little fuel and has no warm, glowing planet in its way as it stares out into the universe. The nature of Lagrange points means that this halo orbit is around a location in space that itself contains no mass, a peculiar consequence of the balance of forces at play.

The arrangement of these Lagrange points also illustrates the strangely hierarchical nature of stable orbits and the precipitous decline of gravitational force as the inverse of the square of the distance of separation. The Earth and Moon have their five Lagrange points, spread across about 2 million kilometers, but then the amalgamated object of the "Earth-Moon" as paired with the Sun has *its* five Lagrange points, spread across some 94 million kilometers. This can occur because the Earth and its moon are close together enough and far enough from the Sun that the Sun's varying pull across the span of the Earth-Moon system isn't sufficient to disrupt the local balance of forces. The same kind of nested hierarchy exists for planets and moons, and even for moons of moons, with hints that such things might have once existed around Saturn's moon Iapetus and might still exist for some of the smaller and more obscure satellites of the giant planets.

The farther from the epicenter of the Sun's sloping gravitational pull, the easier it is for planetary hierarchies to exist. Pluto, about 6 billion kilometers away, is paired with the body called Charon, which is about an eighth of Pluto's mass. These two objects orbit around a point between each other, like a lopsided dumbbell, or more technically a binary world or (to add to the confusing profusion of labels) double-dwarf planet. But in orbit around the outside of this waltzing pair are four other moons, or circumbinary satellites, going by the names Styx, Nix, Kerberos, and Hydra, making up a hierarchical underworld kingdom governed by an unseen lord of gravity.

Intriguingly, the lower branches of these orbital hierarchies seem to end somewhere around the realm of moons of moons, and the reason is partly to do with the material nature of these objects. There is only so far that matter like ice or rock and iron can compress itself in normal conditions. This leaves planets and moons comparatively loosely put together and vulnerable to gravitational tides, where one side is pulled on more strongly than the other. Beyond a certain point, those tides just tear everything apart. This limit, known as the Roche limit after the French astronomer Edouard Roche, who first calculated it in 1848 (eleven years before Darwin published *On the Origin of Species*), indicates the closest distance an object can orbit another object before it is shredded to pieces.[8] But moons also need to orbit close enough to their host world to resist the pulls of other bodies in a system and allow that orbit to be stable. For moons of moons, that seems to ensure that stable orbits must be so tight that they flirt with disaster and the Roche limit. One set of orbital rules giveth, and the other set taketh away.

Further orbital oddities can be found very close to some planets, though. Around the Earth, for example, there is the well-known geo-stationary orbit, at an altitude of around 36,000 kilometers, whose period exactly matches Earth's twenty-four-hour day. An artificial satellite placed in this trajectory—aligned with the planetary equator—will seem to sit nearly motionless in the sky, sometimes tracing out a small figure-of-eight pattern. Another special configuration around the Earth is what's termed a sun-synchronous orbit. In this case an object orbits from pole to pole at an altitude of about 700 kilometers, but the plane of that orbit slowly rotates, or precesses, so that it makes a complete turn in precisely one planetary year. This allows us to place satellites, like the Landsat remote sensing spacecraft, to continually pass overhead with exactly the same angle of illumination by the Sun every time, which is very helpful for gauging shadows and topography from space. Some sun-synchronous craft even perpetually track the terminator, the line between night and day, in order to be continually powered by sunlight and yet always peer into the night.

The secret to this kind of shifting orbit lies in the Earth's own shape, which is not a perfect sphere but rather an oblate spheroid that is fatter at the equator. This deviation from perfect three-dimensional symmetry allows an orbit to be found where that equatorial bulge actually provides the uneven gravitational pull to cause the orbital precession. You can also do this around Mars, but Venus is so close to being perfectly spherical that it won't work without periodic help from a rocket.

The diversity of these orbital variants and gravitational highways and byways presents an intimidating landscape for a species bent on extending itself to interplanetary space, heading for what might be a Dispersal. But we've already adapted in many ways to our spread into the solar system. Familiarity with the ebb and flow of planetary location and the demands of space travel has helped us figure out unexpected opportunities for exploiting the tides and currents of gravity, just as we've exploited the tides and currents of Earth's oceans and air.

Perhaps the most famous example of this process had its conception in the spring of 1965, when the aerospace engineer Gary Flandro, then working as an intern at the Jet Propulsion Laboratory in Pasadena, was investigating options for sending spacecraft farther than ever before, all the way to the outer planets: Jupiter, Saturn, Uranus, and Neptune.[9] This seemed like an overwhelmingly tall order. Even the first port of call, Jupiter, is nearly three and a half times farther from the Sun than Mars, and Neptune is twenty times farther. With the rockets of the 1960s, it was looking like it would take a thirty- to forty-year trip to visit these worlds. Whatever initial velocity we could muster to fling a probe from the Earth into the outer solar system would be eroded away as it climbed out from the Sun's gravitational pull. This would lead to a painful slowdown, lengthening the journey to decades in an era where it couldn't be guaranteed that a spacecraft would last more than a few weeks or months at best.

But in 1961, Michael Minovitch, another young scientist and a predecessor of Flandro, had begun to use a computer (still an expensive and novel tool at the time) to plot the movement of the solar system's planets and the effects of their gravity on a spacecraft moving across interplanetary distances. This work had explored a novel possibility that suggested that spacecraft could exploit Newton's physics rather than be bowed by it. In fact, this idea had been scattered across the scientific literature as far back as the 1920s (in some cases in Soviet journals that would not see the light of day in the rest of the world for decades). It appeared that a probe could, in principle, fly by a massive planet and deliberately allow the planet's gravitational pull to briefly accelerate it in the direction of the planet's orbit around the Sun. In doing so the probe would steal a small piece of the planet's orbital momentum to boost its own velocity—in effect, hitching a gravitational ride without stopping at the planet. Because a spacecraft is tiny and a planet is big, what was a miniscule loss of momentum for the planet (causing less than a trillionth of a trillionth of a kilometer per second change in its velocity) could be a large gain for the spacecraft, throwing it onward with velocity gains of tens of kilometers per second while also adjusting its trajectory without expending any fuel at all.

A much simpler version of this kind of gravity assist had already been used by the Soviet Luna 3 probe, as I discussed. But now, for Minovitch and Flandro, armed with enough computing power to design precise flyby trajectories, and with the tantalizing promise of deep interplanetary exploration, this was an idea ripe for testing. With gravity assists in mind and eager for possible ways to reach the outer solar system, Flandro started charting out the future configurations of the planets, looking for a window of opportunity.

What he realized was that in just ten years from then, in the mid-1970s, all of the giant worlds—Jupiter, Saturn, Uranus, and Neptune—would be positioned on the same side of the solar system, arranged along what was, in effect, a great spiral pathway in space and time. This configuration wouldn't happen again for another 176 years, but if a probe could be sent in time and if it used each planet to fling itself onward to the next, it could visit each world within just 12 years of being launched.

Just getting a probe out as far as Jupiter was a highly uncertain proposition in the 1960s, though. At that distance, with sunlight some twenty-five times fainter than at the Earth and with solar-panel technology still in its infancy and not terribly efficient, such a spacecraft would need an alternate power source. The best solution was to use fearsomely hot packs of radioactive plutonium-238 oxide, whose heat would drive the generation of an electrical current in a sandwich of different conducting materials called a thermocouple. Outer-solar-system probes would also have to carry powerful communication systems and enough fuel to maneuver and orient themselves in the farthest reaches of the solar system. They would also need to be built in a way that could allow adaptation to unforeseen circumstances, whether through the redundancy of systems to compensate for things failing or to adjust to the unknown places that they would be visiting.

The story of how plans were made to first explore the outer solar system is a fantastic one, full of twists and turns and messy human complications, and I'm not going to repeat all that has been written about it before. But the possibility of an orbital grand tour of our solar system seemed too good to pass up, and a hard battle of scientific wills and political and budgetary wrangling began not long after Flandro's critical insight.[10] As a first step toward learning more about the risks of such a deep-space venture, the United States built two nearly identical probes called Pioneer 10 and 11 that launched in 1972 and 1973, respectively. Both would fly by Jupiter, and Pioneer 11 would go on to Saturn before heading to the empty space of the outer solar system like its sister craft. Each performed a gravity-assist and redirection maneuver at Jupiter, with Pioneer 10 being the first-ever spacecraft to slingshot around a giant planet, reaching an eye-watering 132,000 kilometers per hour at its closest approach to Jupiter before heading out to interplanetary space.

These encounters with Jupiter were spectacular and showed that deep-space exploration could be accomplished this way. During this same period, NASA gathered a group of scientists who put forward a refreshed plan for a complete grand tour using two spacecraft. One would visit Jupiter, Uranus, and Neptune; the other would visit Jupiter, Saturn, and Pluto—with Jupiter's enormous mass and orbital speed providing a critical gravity assist for both probes. With these separate machines, the entire tour could be completed in only 7.5 years. The only downside was a cost close to $1 billion, a colossal sum for a scientific mission at the time and more than political and budgetary pressures could stomach.

Luckily, the experience of the Pioneer probes had provided scientists with a wealth of knowledge about how to design an even better set of interplanetary spacecraft, as well as how to strategically maneuver through the mysteries of politics and budgets. With some quick legwork a new, somewhat cheaper mission plan was created that could still launch on time to visit all of the giant planets (but not Pluto) and survive in the ever-darkening depths of interplanetary space. In 1977 this became, with a renaming, the Voyager mission, with two separate but identical probes launched in August and September of that same year.

What a voyage for the Voyagers it has been: akin to the lengthy oceanic explorations of ships like the *Beagle* but also something new, a kind of generational trip that alters our conceptions of causality and scale, foreshadowing the possibilities of Dispersal. By 1979, two years after launch, both spacecraft encountered Jupiter and some of its great moons, like the intensely volcanic Io and icy Europa. Meanwhile, back on the Earth, the USSR was busy invading Afghanistan, Margaret Thatcher was being elected as prime minister of the UK, and the core of the Three Mile Island nuclear reactor was undergoing partial meltdown, all to the tune of Pink Floyd's *The Wall*. These terrestrial issues may have occupied human thoughts, but for the Voyagers they were an ever-receding concern as the probes scooted around Jupiter for their gravity assist, redirecting toward Saturn with a gain in velocity of between 10 and 20 kilometers per second—an unearthly boost.

In 1980 through 1981, the Voyagers hove into view of Saturn, gaining the closest looks that anyone had ever had at its moons, like the hydrocarbon-doused Titan, and revealing more of the complex dynamics of the gas giant's eponymous ring system. Back on the Earth, Mount St. Helens had erupted, and John Lennon had been killed. NASA's space shuttle had made its maiden voyage, and Indiana Jones wisecracked and whip-cracked his way onto cinema screens in *Raiders of the Lost Ark*. Saturn was also a turning point for the

Voyagers because this was where the probes parted ways. Voyager 1 would never again pass near another planet, instead being flung outward on its next phase through the depths of interplanetary space toward the mysteries of interstellar space.

By contrast, Voyager 2 had two further, critical visits to make. It first encountered Uranus and its moons in 1986, and then, with another gravity assist, it encountered distant, chilled Neptune in 1989, closing out a decade in which—back on Earth—the Berlin Wall came down, Microsoft introduced a newfangled word processor called Word, *The Simpsons* debuted, and the Galileo spacecraft was launched to eventually reach and orbit Jupiter by 1995. After its final gravity assist, redirecting from its passage near Neptune, Voyager 2 also headed outward, away from the Sun.

In 1990 Voyager 1 was famously commanded to pivot and take a series of images across the system it was leaving behind, a family portrait. By this time, the probe wasn't just beyond the orbit of Neptune; it had also climbed steadily up and away from the plane of the solar system—the disklike geometry in which all the major planets orbit. From this unique vantage point, its pre-1977 cameras used photosensitive cathode-ray tubes and a mechanical eight-color filter wheel to take sixty frames that captured the Sun—now 1,600 times fainter than when viewed from the Earth—and the reflected light of all the large planets, except for Mercury and Mars. These data were stored in a digital tape recorder, and several months later, that information was finally and painfully slowly transmitted back to the mission scientists.

In among these images was a carefully choreographed snapshot that contained the Earth in a single pixel, a tiny speck of pale-blue reflected light, now immortalized as the "Pale Blue Dot": a profoundly meaningful point of light that represents the truly humbling condition of humanity and the condition of life in the universe. It is an image that has now entered the popular lexicon, helped along by the planetary scientist Carl Sagan's supremely eloquent and devastatingly pithy description.[11] But although this dot focused our attention on the pitiful insignificance of our parochial attitudes and misplaced energies when compared to cosmic reality, there was another, equally important, side to the accomplishment.

The moment that Voyager 1 swung around and began capturing these family pictures was both a closing parenthesis and a critical forward marker for a special journey of adaptation and intention that stretched back past people like Laplace and du Châtelet, past Newton, Kepler, Galileo, and so many others in Europe, the Middle East, and Asia, and even past the rise of modern humans, all the way to some unknown moment four billion years ago when life began to invent itself in the chemical stew of a young world. And if Voyager's camera had been sensitive enough, it would have seen this pale blue dot now shedding a cloud of even tinier dots in an evaporative puff of rockets and glinting metals—the surefire signature of life growing past a planetary enclosure.

Living systems follow an endless assignment to gather, process, and respond to information about their environment, both amplifying and modifying the selective pressures of survival by allowing organisms to make irreversible and extraordinary decisions. The pale blue dot was one part of that ongoing assignment for life on Earth and a confirmation of sorts of life's gradual awareness and expectations about the configuration of worlds and the environment that produced it. We may not have fully appreciated it, but we've been living in the aftermath of that moment and all of the other moments of the past decades of space exploration, and we are already walking some very new pathways.

Together with the now incommunicado probes of Pioneer 10 and 11, the Voyagers have gone on to become the first devices from the Earth to pass through the distorted bubble of flowing particles that is the Sun's domain.[12] This is the outermost stellar atmosphere, which distinguishes this local volume of the cosmos from the surrounding contents of interstellar space. In 2012, Voyager 1 was the first to pass through the interface zone of that bubble known as the heliopause, at a distance 121 times farther from the Sun than the Earth orbits, some 18 billion kilometers. Then, in 2018, at a similar distance, so too did Voyager 2. That initial passage into interstellar space was registered in a wonderful and lucky way. Many months before, a solar storm of electrons and protons had erupted outward. By the time it washed across Voyager 1, it acted like a roll of distant thunder echoing off a nearby mountain hidden in cloud. It caused a measurable disturbance in the probe's surroundings that enabled Voyager 1's instruments to sense that this environment had changed and was now a part of interstellar space.

At the distance the probes are at today, electromagnetic radiation—light itself—takes more than eighteen hours to reach Voyager 2 from the Earth and more than twenty-two hours to reach Voyager 1, and both machines are—as I write this—still receiving and sending signals. Forty-five years of history have unfurled on the Earth since these two launched. Human actions have systematically increased the concentration of carbon dioxide in the planet's atmosphere, taking it from some 330 parts per million to over 420 parts per million today. And in 1977 there were 4.2 billion humans alive, but now there are almost 8 billion.

In those forty-five years there has been an exponential growth of rockets and spacecraft lofting into Earth's orbit and well beyond. Many more scientific missions have followed the example of the Voyagers by catching gravity assists. In fact, most missions to anywhere beyond Earth's immediate vicinity are now constructed around the use of planetary encounters. We've gotten so good at this that missions like the European Space Agency's Rosetta probe exploited multiple flybys of Earth and Mars, spanning a period of ten years, scooting to within 250 kilometers of Mars between its first and second gravity assists by the Earth, and before its third such maneuver.

All of this was used to push the spacecraft out to the asteroid belt, where it encountered two large asteroids and, eventually, its goal of the cometary body 67P/Churymunov-Gerasimenko in 2014, where it entered orbit around this 10-billion-ton, gnarled lump of primitive material.

An even more circuitous route is planned for the European Space Agency's Jupiter Icy Moons Explorer (JUICE), launched in 2023.[13] This mission will first complete an orbit of the Sun that closely matches Earth's. It will get a gravity assist from the Earth that sends the probe to Venus, then back to Earth and on to Mars, then back again to Earth for a final flyby that will send the spacecraft outward to intercept Jupiter some three years later and nearly eight years after its launch. It brings new meaning to the refrain "Are we there yet?," but there is an excellent reason for this strategy because it saves on launch costs and on mass so that the scientific payload can be as large and as useful as possible.

Those payloads and the designs of spacecraft continue to evolve. It's a long way today from the wet-lab film canisters of Luna 3 or the generation of 1960s and 1970s spy satellites that would sometimes spit their exposed images into special return vehicles. Those little pods would have to reenter the Earth's atmosphere to be retrieved, sometimes in midair. Solid-state microelectronics have replaced such things, as well as the magnetic tape recorders of probes such as the Voyagers. Miniaturization and technological advancements have offered ways to pack more and more instrumentation into commercial or scientific spacecraft and have helped fuel a prolifera- tion of new forms and approaches to robotic spacecraft, as well as human-carrying vehicles.

Massive satellites and probes are still produced for flagship mis- sions, though. A craft like JUICE, heading to Jupiter, is a 2.5-metric- ton behemoth with 3 metric tons of propellant on board. It has advanced solar panels spanning 85 square meters that are capable of generating nearly 1,000 watts of power even though the Sun appears some twenty-five times fainter at the distance of Jupiter. It also carries eleven complex scientific instruments, from cameras to laser altimeters and radars, as well as a sophisticated communications system for pumping data back to Earth's ground stations at a rate of 2 gigabytes per day.

There have been advances in tiny spacecraft as well, including the so-called cubesat, which is built around a fixed basic unit, or form factor, and is a 10-centimeter-wide cube with a mass of around 2 kilograms that can be connected to others.[14] These compact machines easily piggyback as ride-alongs with other missions. Small, legal hitchhikers can get to Earth orbit or to interplanetary space, as in the case of the 2018 Insight mission to Mars, where a pair of six-cube-large cubesats performed as data relays while Insight landed on the planet. The most critical and extraordinary thing about cubesats is that their availability and low cost, with some three hundred launches per year, have enabled more and more nations to, in effect, start their own space-exploration programs.

Many of these programs use the vantage point of Earth orbit and what it provides for communications or for remote sensing of a nation's lands and resources. But many are for scientific questions, striving for some of the same kinds of insights and revelations that probes like the Voyagers have given us. In space, or on the surfaces of other worlds, are opportunities to make the most exquisite measurements of environmental characteristics or to sniff around for clues to deep questions about the origins of planets and the existence of life itself. The pristine nature of the interplanetary vacuum or a planetary surface makes it possible to detect and measure phenomena and material compositions with extraordinary precision. Consequently, we've learned to be very, very good at doing this. The kinds of scientific instruments we send out into the solar system can see and sense in multiple wavelengths of light, and they can taste individual atoms and molecules that exist in concentrations where one in a million is an abundance.

Rather unexpectedly, these experiences have also taught us that the only way to gather the data we want is to tread extraordinarily lightly. When a piece of machinery goes into space, it starts shedding atoms and molecules into the vacuum, and this can confound attempts to examine indigenous materials. Conductive structures and electronics can also generate radio waves and magnetic fields that are similarly polluting. Even intensely cleaned and sterilized equipment can still carry terrestrial microorganisms. For spacecraft, that's a problem exacerbated by particularly hardy species of microbes whose idea of a perfect vacation is inside a spacecraft-assembly clean room, where competing organisms have been eliminated. If these clean-room survivors are transferred onto spacecraft, they might confound attempts to look for signs of life elsewhere.

Consequently, spacecraft are no longer just bolted and glued together with handy pieces. We train experts to learn the behavior of every millimeter of material and bonding agent that goes into space so that there are as few polluting factors as possible. A robotic craft becomes a varied landscape to study and understand in its own right, where the wafting gases from one area might impinge on the delicate instruments of another. Decisions are made on such minutiae as whether the thin edge of a solar panel needs to be sealed tightly in order to avoid the risk of contaminating a measurement of gases that are being expunged by an icy moon around Saturn or Jupiter. The mirror coatings of telescopes are examined to understand what spectral ghosts they might impose on delicate measurements of light from the distant cosmos. It is a painful, meticulous process with few parallels to anything else that we do on Earth, where we live in a seething swamp of filth.

In 1958, when plans were being made to set out across our tiny patch of universe, none of these things were really on the agenda.

In retrospect, it is clear that the movement of life beyond the Earth is both enormously complex and deeply *meaningful* as a physical extension of human intention and creativity. A rocket or spacecraft is a piece of intricate art that reshapes our conceptions of form and function. It can be a cubist masterpiece that happens to combine reactive chemistry and seat-of-your-pants aeronautics in splendid and hair-raising synchrony. Equally, interplanetary travel is as non-intuitive as sailing into the wind, and it comes with at least as many unexpected opportunities.

To follow our compulsion for this thing that we call space exploration, we've invented whole new fields of expertise in areas like orbital mechanics, spaceflight mechanics, space communications, and "astro-navigation," a specialty that has become a near art form for a small cohort of highly specialized mathematical wizards across the globe like Gary Flandro. To precisely intercept planets or other vehicles is a feat akin to deftly shooting at fleas across thousands of miles of distance. The emergence of these fields of endeavor, just as with the advancements in navigation enabled by the *Beagle*'s chronometers and that era's expert mapmakers, is a part of an evolutionary moment in which we're a pioneer organism flooding a previously untouched and untouchable landscape.

In that landscape, few other worlds have captured as much attention, rightly or wrongly, as the next place that we'll discuss. And few other places shine as harsh a spotlight onto the motivations and possibilities for human space exploration as this strange, reddish rocky planet.

— 6 —

THE RED SIREN

How great would be the desire in every admirer of nature to
behold, if such were possible, the scenery of another planet!
—Charles Darwin,
The Voyage of the Beagle

The campus of the Jet Propulsion Laboratory is squeezed up
against a set of steep hills close to Pasadena, from where, if you're
able to face the climb, you can catch a glimpse of the sprawling con-
urbation of Greater Los Angeles. These hills are part of the Pasadena
orogeny, a surge of crumpled mountain building that started millions
of years ago when the Earth's great crustal North American Plate
shoved into the Pacific Plate: the kind of multimillion-year violence
that regularly happens to a rocky planet like ours and is unappreciated
by most of the scraps of short-lived biology that occupy its surface.

If you drive to the laboratory, you'll be funneled from the artificial concretions of the San Fernando Freeway onto a modest road that
passes scrubby parkland, a riding stable, and a high school that looks
like it was lifted out of Disney's Tomorrowland as a white-stucco,
1960s vision of what the future was going to be. Tapered columns

and curious geometric layers are everywhere. The laboratory itself—
JPL, as it's almost universally called—emerges between the trees and
shrubs as an updated version of this projection, with concrete giving
way to the smoked-glass and steel exteriors of the early 1970s.

Once you're through the security entrance, a short walk up a few
flights of steps will take you to the Spacecraft Assembly Facility and
its great clean room, High Bay 1. This cavernous, brightly lit, and
heavily air-conditioned chamber has seen generation after generation
of spacecraft come and go: the Mariner missions of the 1960s, the
Viking orbiters, and the Voyagers, as well as the Galileo mission to
Jupiter and the Cassini probe to Saturn.[1] They were all assembled and
tested, and tested again, in this facility before being squeezed out of
its huge doors and taken off to be rocketed into space.

But just before you walk through to High Bay 1's glass-enclosed
viewing gallery, there's a discreetly placed installation on the walls
that's easy to overlook. At its center is a framed artwork, a study in
tones of rust and beige that might have emerged from a collaboration
between Rothko and Rauschenberg, covered with spidery annotations
and lines that appear to form crosshairs over the image. Look closer,
and you'll see that the entire picture is made of dozens upon dozens
of vertical strips of paper, all neatly glued down. It's an intriguing
and attractive construction, but any aesthetics are quite inadvertent
because this picture is actually a color representation of the very first
close-up view of the surface of Mars.

In July 1965 the Mariner 4 spacecraft—built here in High Bay 1—
swept by Mars at a distance of around 10,000 kilometers and trained its
scanning television camera on the planet before transmitting the data
back to Earth at a painfully slow 8.3 bits per second. As these data trick-
led back, they were relayed to JPL in teletyped columns of numbers that
corresponded to the electronically assigned brightness of the image. The
computers that processed this information were slow, and the scientists
and engineers were supremely anxious to know that the mission had

worked. Rather than waiting in agony, they quickly devised a scheme to build their own picture in a way that would've made their kindergarten teachers proud. The paper teletypes were cut into skinny columns, hand colored with a set of artist's pastels (now also preserved in a frame on the wall), and pasted together to make a map. If you look closely enough, you can see the faded teletyped numbers beneath the dusty pastel layers.[2]

There's little real detail in this chart; it's a vista of suggestive tones, without much by way of notable markings. Altogether, it represents an obliquely viewed swath of the Martian surface and planetary "limb," or horizon, that is only 300 kilometers across. It's a TV glimpse of a world passing by like some lingering residue from late-night channel surfing. Yet although it's astoundingly primitive by modern standards, this picture was among the first truly digital images to be formed anywhere—on the Earth or at another planet. All in all, including this image, Mariner 4 captured a total of twenty-two glimpses of the surface—a trail of snapshots lined up as the spacecraft raced by at 25,000 kilometers per hour.

Despite their relative crudeness, these pictures, along with the rest of Mariner 4's scientific measurements, were a revelation. Mars seemed to be cratered and desolate in appearance. It had no global magnetic field, and it had the thinnest of atmospheres. There was absolutely no evidence of liquid water on its planet-encircling landmass. It appeared utterly dead—a cold, dry desert world. With the capture of these vistas, centuries of speculations and wishful thinking about our Martian neighbor were, almost literally, swept away in a cloud of dust. There was no alien civilization, no network of canals or cities, and no sign of a living thing. This was such a profound discovery that when members of the Mariner 4 team were invited to the White House in recognition of their accomplishment, President Lyndon B. Johnson (or his speechwriter) was moved to express the fact that "it may just be that life, as we know it, is more unique than many have thought."[3]

Yet if anything, this discovery only created a greater urgency for us to understand Mars. Why was our rocky neighbor inhospitable—how could this be? Even Mariner 4's pioneering measurements didn't quite convince everyone. After all, the images and data it sent back were just a peek across one stripe of the planet's surface, and there was still a possibility of regional differences and further mysteries on Mars. The thirst to visit the red planet was unquenched, and 1969 saw the ambitious twin launch of Mariners 6 and 7. These identical probes were sent to fly by Mars within days of each other, doing so just a week after Apollo 11 had placed humans on the Moon.

The pair of Mariners still captured images of only parts of Mars—missing many of the major geographic features that we now know—but they determined that Mars's thin atmosphere was mostly carbon dioxide and saw that there was evidence of small amounts of water in the environment and more frozen at the poles. Mars wasn't a lost cause yet. For a brief time, the United States even came up with a vision of quickly using one of the massive Saturn V Moon rockets to take a set of two robotic orbiters and two robotic life-seeking landers to Mars, followed by humans.

Before and after Mariner 4, there have been more than fifty efforts to send probes to Mars, a total second only to the number of lunar missions.[4] Earlier, in 1960, the Soviet Union had made humanity's very first attempt to reach Mars by launching a 650-kilogram, solar-powered spacecraft to scope out the interplanetary journey. But this ended rather ignominiously with a cascade of problems during launch and, instead of heading toward Mars, the mission reached all of 120 kilometers above Earth's surface before disintegrating across Siberia. The next five attempts to reach Mars, mostly by the Soviets, were equally ill-fated.

Between 1964 and 1971, no fewer than thirteen new probes were built, with eight of these suffering from absolute or partial failure, either at launch or somewhere along the lonely path through interplanetary space to the hazardous encounter with Mars. Those failures included Mariner 3, a near twin of Mariner 4 that had, just a few weeks earlier, been unable to open its solar panels after part of the launch-containment structure—called the fairing—had crumpled and damaged things. Even Mariner 4, despite its eventual success, was a hair-raising mission. It had to deal with temperamental star trackers and tape recorders, uncertain orbital positions, and a complicated sequence of actions that all needed to succeed just to ensure that the spacecraft pointed in the right direction as it flew by Mars. It was a bit like throwing an exceedingly complex bundle of machinery out of an aircraft and hanging on for dear life to operate it.

One of the most intense periods of exploration was in May 1971, when the calendar of favorable Mars transfer orbits resulted in the launch of five missions targeting Mars. Among these were the Soviet Mars 2 and Mars 3 projects, which each carried an extraordinarily ambitious payload of an orbiter, a lander, and a small rover. The Mars 2 probe launched from the Earth on May 19, 1971; Mars 3 launched on May 28; and both arrived at Mars within days of each other some six months later. For Mars 2, the orbital insertion around Mars worked, but the lander and its rover crashed into the surface of the planet. When Mars 3 arrived, it too got into orbit, and then, using a sequence of aerodynamic decelerations, parachutes, and retro-rockets, its lander made a "soft" touchdown. After that, though, it transmitted only twenty seconds of data, including a partial but seemingly blank image, before falling quiet.

The little Soviet rover on Mars 3, like the one on unlucky Mars 2, was a remarkable thing called PrOP-M.[5] This machine was a 4.5-kilogram box sitting on a pair of skis and attached to the lander by a 15-meter electrical umbilical cord. By "walking" the skis, it could move forward, and by having the skis move in opposite directions, it could turn. Because of the signal travel time from the Earth, the rover even had a primitive version of autonomy, with a pair of bumper-like mechanical sensors that would trigger the rover to turn if they encountered a rock or other obstacle. Sadly, with the failure of the rest of the lander we won't ever know whether PrOP-M got to shuffle across the surface of Mars until someone goes and finds where Mars 3 landed and takes a look.

Exactly why Mars 3 went silent has never been determined, but one suspicion is that Mars itself was the culprit. Just before Mars 2 and 3 arrived, the United States had, by two weeks, beaten the Soviet Union to successfully orbit the first-ever spacecraft around another planet with the Mariner 9 probe.[6] However, when Mariner 9 arrived at Mars on November 14, 1971, it found a world almost entirely shrouded in a global dust storm. This storm was so severe that Mariner 9 literally couldn't see the planetary surface, so the probe was urgently reprogrammed to wait things out before attempting to perform most of its scientific observations. It seems quite likely that this very same storm, still raging when Mars 2 and 3 showed up, could have caused serious problems for the Mars 3 lander. Neither of those Soviet missions had the capacity to be reprogrammed for these unexpected conditions.

As it was, Mariner 9 had to wait about two months in orbit before the Martian skies cleared enough for it to start carrying out the majority of its work. That work included the first thorough photographic mapping of Mars's surface, which revealed that Mars had plenty more surprises to share. Among Mariner 9's discoveries were the first images of the vast geological form of Olympus Mons—already poking suggestively through the dust-filled skies at the probe's arrival. This 22-kilometer-high shield volcano is the largest such geographical feature in the solar system, and it spreads across an area similar in size to the country of Poland. Olympus Mons is so massive that it actually depresses Mars's crust, forming a 2-kilometer-deep moat around itself. Furthermore, just a thousand kilometers to the south is a chain of three other enormous volcanic mountains, the Tharsis Montes, that are nearly as tall as Olympus Mons itself.

To the southeast of these volcanic zones lies the monstrous equatorial Valles Marineris: a gouge-like canyon system that is as long as the United States is wide, has a depth as great as 7 kilometers, and is thought to be a result of giant rift faults, where the planetary lithosphere has been pulled apart. Mariner 9 also found what appeared to be enormous flood channels and other tectonic features on Mars, making it clear to geologists that this was a world with an exceedingly complex history. In fact, the orbital maps revealed that Mars is a world literally divided in two. This Martian dichotomy, as it has become known, is a striking divide between north and south.[7] Much of the northern hemisphere is wrapped in lowlands, to the extent that a third of northern Mars is lower in elevation by 1 to 3 kilometers and is flat and monotonous, whereas the southern two-thirds of the planet harbors its giant volcanoes and canyons, as well as rugged highlands. We don't know exactly what created this dichotomy, but one hypothesis is that it's the result of a cataclysmic planetary collision in the same vein as the event that is thought to have created Earth's moon. All of these discoveries were a clear lesson that in order to understand Mars as it is today, we'd have to understand its past.

The rapid pace of launches to Mars continued after Mariner 9. In 1973 the Soviet Union launched Mars 4, 5, 6, and 7, and although none of their landers succeeded, they gathered more and more data from orbit. These probes also encountered other hazards of interplanetary exploration. Mars 5, for instance, seems to have been hit by a micrometeoroid as it was entering orbit around Mars, causing its pressurized compartment of sensitive electronics to leak and eventually fail. Mars 6, which was primarily a lander, lost all contact just as it was about to fire up its retro-rockets after parachuting through the atmosphere. The Mars 7 lander missed its positional window for entering the atmosphere and instead sailed off into orbit around the Sun—a fault attributed to bad transistors in the spacecraft's electronics.

Then, two years later, came one of the most ambitious and influential projects in the history of space exploration and space science. In late August and early September 1975, the United States launched two nearly identical missions, each consisting of an orbiter for Mars and a large, sophisticated lander. The Viking 1 and 2 projects marked a special moment, both for the amount and quality of scientific data they would obtain and for the instruments on the two landers that were designed explicitly to search for direct evidence of life on another world.[8]

The spacecraft arrived at Mars about a month apart in 1976. A tense few weeks passed in orbit while the probes surveyed Mars to find safe landing areas. But each lander was finally detached and de-orbited, slowing in the atmosphere in a hair-raising hypersonic entry, before deploying parachutes and radar-controlled hydrazine retro-rockets to slow their descent. The two landers set down on Mars at sites about 6,500 kilometers apart in the easier terrain of the northern hemisphere. These stationary Vikings were hefty creatures, each the size of a small car and weighing in at nearly 600 kilograms. Both were powered by 70 watts of electricity provided by plutonium-fueled radioisotope thermoelectric generators. They were crammed with state-of-the-art technology and obtained the first color digital images from the surface of Mars. They each had a robotic arm for digging and scooping Martian soil, along with gas and X-ray spectrometers for sniffing at the composition of materials. And they each carried a 16-kilogram package of biological experiments designed to search directly for signs of life in the Martian soil.

That search for life gave wholly confounding results, with four different experiments yielding a perplexing mix of positive and negative findings. Today, most scientists consider these results a consequence of what was, at the time, unknown and unanticipated soil chemistry.[9] Mars's surface soil can contain highly reactive nonbiological oxidizing compounds that would have caused havoc inside the landers' miniature labs. If there was or had ever been life at the Viking landing sites, it wasn't likely to show up easily in these conditions and with these tests.

These results were a traumatic lesson, and as priorities and projects in space exploration shifted and evolved, there was a gap of some twelve years before any nations launched for Mars again. When they did, starting with the Soviets in 1988 and their Phobos 1 mission to probe Mars's moons, failure seemed to be almost the only option. First, Phobos 1 was prematurely shut down—likely because of a single missing hyphen in a transmitted command code. Then Phobos 2 managed to take thirty-seven images of its namesake moon before a computer issue shut it down just two months into its time at Mars. Next, in 1992 the United States' Mars Observer was lost three days before it was supposed to enter Mars's orbit, probably as a result of a rupture in its propulsion system. In 1996 there was a turn of fortune when the Mars Global Surveyor worked just fine (and continued to do so for an entire decade in orbit). But then the large and ambitious Russian Mars 96 probe failed very ignominiously, falling into the ocean off Chile after three orbits of the Earth, although the United States' Mars Pathfinder mission successfully deployed the first operational surface rover, little Sojourner, also in 1996.

In 1998 Japan's Nozomi orbiter had an electrical failure, and the United States' Mars Climate Orbiter burned up in the Martian atmosphere because of a software error that meant one internal system expected metric units of measurement while another expected imperial units. In 1999 the Mars Polar Lander crash-landed, and its two Deep Space probes failed. In 2001 Mars Odyssey successfully kicked off its mission to map Mars's mineralogical composition and was still going in 2025 after more than a hundred thousand orbits, with only one failed instrument, caused by a solar storm in 2003. And in 2003 the European Space Agency's Mars Express went into orbit and was also still working as of 2024, although the United Kingdom's Beagle 2 probe—carried with this mission—made a disastrously violent landing on the surface.

Since that time, the other successful missions have been the Spirit and Opportunity rovers, the Mars Reconnaissance Orbiter, the Phoenix lander, the Mars Science Laboratory (a.k.a. the Curiosity rover), India's Mars Orbiter mission, the MAVEN orbiter, the ExoMars Trace Gas Orbiter, the Insight lander, the Emirates Mars mission, China's Tianwen-1 (including the Zhurong rover), and Mars 2020 and its Perseverance rover and Ingenuity helicopter. Some of these missions have pioneered highly inventive ways to reach the surface of Mars intact. This is no small challenge: Mars is massive enough to have a significant gravity well to accelerate falling objects, but its atmosphere is so thin that many familiar techniques of slowing down from orbital velocities to gentle landings are seriously hampered. The Pathfinder, Spirit, and Opportunity missions all used an ingenious approach that combined parachutes for slowing in the atmosphere, a set of small retro-rockets, and, for the actual landing, a cocoon of large, rapidly inflated airbags so that the probes bounced for hundreds of meters across the landscape until coming to a cushioned rest. Even that audacious technique was one-upped, though, when the two more massive rovers Curiosity and Perseverance (in 2012 and 2021, respectively) were actually lowered from the underneath of a flying, hovering, autonomous "sky crane" directly to the surface on 7-meter-long cables.

At this moment there are eleven operational robotic orbiters or surface devices at Mars, with arrival dates going back to 2001. It's an obvious statement to make, but Mars is the first machine-occupied world in the solar system. That isn't just frivolous hyperbole; it's a profoundly important statement about the ongoing, demonstrable evolution of our species' presence in the cosmos and about the dispersal of terrestrial life's influence across the solar system. In that context, it's valid to see all of those early flights to Mars—filled with failures as well as revelatory discoveries about the nature of another world—as a series of natural experiments carried out as life on Earth tried to gain a foothold in a new landscape. And one outcome of those experiments is that the first new Martians in at least a billion years are not biological; they are robotic.

From a cosmic perspective we could even say that Mars is experiencing an "origin" event with its occupation by these complex devices—akin to an origin of biological life. It is possible that this is the very first origin event on Mars, but it could be the second or even one of many. In layers of minerals and sculpted landforms, the intricately detailed chemistry and elemental mix of Martian rocks and soil are clues to a world that has been many things in the past. That history is distinct enough that, just as we've done for the Earth, we've assigned names to the geological and climatological periods on Mars.[10]

For example, from 4.1 billion to 3.7 billion years ago Mars was in the Noachian period—named for an ancient highland region in the southern hemisphere, the Noachis Terra or "Land of Noah." We think that during the Noachian, Mars had active and extensive volcanism, and it was being bombarded periodically by asteroid material that was rich in water and other chemicals. With a thicker ancient atmosphere raising Mars's surface pressure and temperature, it is likely the planet had persistent liquid surface water and at least the possibility of its own kind of Great Flood. That water was in lakes and maybe even in shallow seas with outflows that etched their river patterns into the landscape we see today, leaving behind clays and waterworn rocks.

The next period was the Hesperian, from roughly 3.7 to 2.9 billion years ago (these age ranges are still quite uncertain), which marked a time of drastic transition on Mars. At the start of the Hesperian, water still flowed, carving great channels here and there, and volcanoes still formed and erupted—including Olympus Mons. Many of those volcanoes appear to have pumped gases rich in sulfur into the environment, causing acidic rains and a chemical erosion of minerals that are reflected in the compositions of Mars's surface today. Massive asteroid impacts were still punctuating the timeline during this period, each with the potential to expunge huge deposits of subsurface water, creating cataclysmic but geologically short-lived floods. It was a period of episodic warmth and wetness. Yet Mars was also undergoing a profound shift toward its future state. With every passing century the planet was becoming a little more frozen, a little more arid and airless. But this was such a gradual change that had life existed (and it may have), it wouldn't have been easy for life to recognize that these things were happening.

From 2.9 billion years ago to now, Mars has been in the rather optimistically named Amazonian period. This cold, dry period is named after another region on Mars, the dusty, smooth volcanic plain of Amazonis Planitia. That region is one of the geologically youngest areas on the planet and has more than a passing resemblance to parts of Iceland on Earth. But even though the Amazonian period extends all the way in time to today's seemingly desolate and dull conditions, it has also seen episodes of startling volcanism, along with all of the environmental enrichment that comes along with that. Those episodes included more activity from Olympus Mons, which may have last erupted some 25 million years ago.

There have also been many Martian ice ages during the Amazonian, driven by the planet's large polar wobble that shifts its spin axis back and forth by tens of degrees over roughly hundred-thousand-year periods and by as much as a chaotic sixty degrees over several millions of years. These movements are brought about by the perturbing interactions between Mars's spinning mass and the gravitational pulls of worlds like giant Jupiter. In fact, they are precisely the kinds of changes that Mary Somerville helped us understand by studying the works of Laplace and Lagrange on planetary orbits. Such wobbles in the planet's orientation drive radical variations in solar illumination and climate, allowing glaciation almost down to the Martian equator at times.

These quirks of planetary history have also left modern Mars with an axial tilt of 25 degrees and a day length of 24 hours and 40 minutes, each disconcertingly close to Earth's configuration. Both are coincidences of our times, though, brought about by the polar wobbles and Mars's formation history. The gravitational pulls of other worlds also cause the shape of Mars's orbit to change over time. Today, its elliptical path brings the planet 17 percent closer to the Sun at perihelion than at its farthest (aphelion) point—a difference five times larger than for the Earth. Combined with the axial tilt, the subsequent changes in surface illumination of Mars drive very strong seasonal changes that are not ameliorated as they are on Earth, where oceanic thermal inertia acts like a storage heater to smooth out the variations. Mars's polar caps are mostly water ice, but they get so cold in their winters that atmospheric carbon dioxide freezes out into an additional solid surface layer that is meters in depth. The southern pole remains cold enough that it also has a permanent carbon dioxide layer. Those seasonal carbon dioxide deposits vary in size by trillions of tons as summer comes and goes for the northern and southern hemispheres. In fact, the entire atmosphere increases and decreases in density as these polar layers of carbon dioxide solidify or evaporate, shifting Mars's atmospheric pressure by as much as a third during a single orbit around the Sun.

We still know only a small fraction of Mars's stories, but insights like these have turned it from a point of mystery in the sky to a rich, diverse, and endlessly fascinating world—a globe of wonders and alternative histories that can help us understand our own world. But what happens next? How does Mars fit into the future of space exploration, and do we warm, wet humans fit anywhere into the future of Mars?

IF YOU COULD STAND outside on Mars, the first thing to strike you would be the silence. The most barren wildernesses on the Earth, whether they're arid and scorching deserts or arctic snowfields, can have moments so quiet that human ears and nerve endings find themselves stunned and in search of something, anything, to register. But replace Earth's atmosphere with one that is a hundred times thinner and made mostly of cold carbon dioxide, and the silence is even more profound. Low-pitched sounds are muffled, and high-pitched sounds are heavily muted for human aural sensitivities, so much so that if there were birds singing on Mars, we probably wouldn't be able to hear them anyway.

The next thing that will strike you (assuming you've not succumbed to asphyxiation) is that you weigh a third of what you're used to, a feeling as if you're in a permanently descending elevator that's accelerating downward just a little too aggressively. It's a very disconcerting effect that permeates every atom in your body and seems at odds with your deepest intuitions, causing a nagging, persistent wrongness. Yet this low gravity can also induce a certain euphoria, something that happened to the Apollo astronauts on the lunar surface when they learned how to "bound" across the landscape even when burdened with the 80-kilogram inertial mass of their spacesuits.

Despite these profound differences, the most bewildering aspect of your experience here on a world some 230 million kilometers from the Sun is how incredibly, deceptively familiar it all feels. There are subtly colored hills and mountains, dry dusty valleys, and wind-eroded rocks in eerie forms that wouldn't seem out of place in the American Southwest or northern Africa. (See Figure 6.1.) Although the Sun is not particularly bright in the sky (it's about 2.3 times dimmer than on Earth), it isn't worse than a rather hazy day on Earth, especially when our terrestrial wildfires or wind-lofted sand colors the air a reddish beige. Mars is not a static landscape, either; skating across your view in the distance are the fleeting swirls of dust devils, even though the breeze is hardly there to be felt. Altogether, millions of these spinning disturbances are known to happen every day across the planet.

If you're standing somewhere in the mid to low latitudes, you'll be able to see the tiny moons Phobos and Deimos tracking across the sky. Phobos moves rapidly, taking just four hours to traverse from horizon to horizon, and it's about a third the size of Earth's moon in appearance. The more distant Deimos moves more slowly and is about 5 percent of our moon's apparent size.

Figure 6.1. The surface of Mars as seen by the Perseverance rover on August 10, 2021. (Credit: NASA/JPL-Caltech/ASU.)

All of these features make Mars sneaky and insidious; it gets inside your head and somehow subverts what is incontrovertibly alien until you delude yourself that it's really quite normal. The human mind is very skillful at adapting our senses and perceptions to environments that are only superficially accommodating. That's true for Earth's oceanic depths, rarefied mountaintops, and achingly beautiful but fearsomely alien worlds separated by millions of kilometers and billions of years. "Lovely Mars," we might say to ourselves; it's not so very different, not so bad. But this is as it starves us of oxygen, disrupts our gravitationally evolved cardiovascular system, corrodes our skin and lungs with powerful oxidants, and punctures our most intimate biomolecules with solar and cosmic-particle radiation while furiously burning us with ultraviolet light from the Sun.

The radiation environment on Mars can come as a surprise, but the planet's tenuous atmosphere, along with the lack of a strong planetary magnetic field, allows its surface to be bombarded with hazards.[11] Just as for astronauts venturing beyond low-Earth orbit, some cosmic-particle radiation smashes into environmental atomic nuclei and releases a cascade of complex secondary radiation that is even more damaging. It is estimated that the average particle-radiation exposure during a year on the surface of Mars can be hundreds of times greater than that on the surface of Earth and could spike if there are large solar storms or chance galactic cosmic-radiation events. Also, although sunlight is overall fainter on Mars's surface than it is on the Earth's, the shortest wavelengths of ultraviolet light—without a filtering atmosphere—are actually more intense. This electromagnetic radiation packs an energetic punch, enough to disassemble exposed organic molecules and to create lots of highly reactive free-radical molecules. These, in turn, damage biological material and even degrade plastics and other compounds.

Those hazards are the ones that we have some approximate insight about, but we have very little knowledge about what would also happen to all of the microbial life that lives in us as our microbiome or coexists in our immediate environments and in our foodstuffs. The same is true of the vast and mysterious ocean of viruses that jostle for survival using the cellular machinery of replication in our cells and in the cells of other organisms. As I've mentioned, some dormant human viruses, like herpes or chicken pox, are seen to reactivate in the environment of space stations, possibly in response to our own body's stresses. The consequences for human health and the longer-term evolution of microbes and viruses on Mars are essentially unknown.

The upshot is that large, complex biological entities from another planet have a lot to worry about on Mars. Conditions here are barely more suitable for humans than conditions on the Moon. For some added perspective, consider the following thought experiment: Imagine that you and your family are dropped onto the top of Mount Everest, where there is not enough oxygen to sustain anyone for longer than a brief time, there is no food, and there is no shelter. Now imagine that you have to stay there indefinitely. Although that sounds like a bad deal, Mars is far worse.

Yet the question of sending humans to Mars, even keeping humans permanently on Mars, has never been more in the headlines than it is today. It's also a question that represents a key element in the profound puzzle of the nature of life's dispersal and survival, a puzzle that is our final destination in this book.

In general, there are three philosophical or, perhaps more accurately, ideological schools of thought on what Mars represents for humans and for life in the solar system. One of these schools of thought is about science, pure and simple, and has existed since those early days of the Mariner and Viking probes. It hinges on the fact that getting a small handful of human explorers to Mars and back would create a wealth of opportunities for scientific discovery. This is in part because of astronauts' capacity to be sophisticated eyes and ears for doing research in real time. But it's also because shuttling humans to

Mars requires more advanced and larger rockets and spacecraft, creating the opportunity to haul far more sophisticated scientific equipment for elaborate measurements and experiments to vastly expand our knowledge of the planet. If we do learn how to explore Mars with humans, we'll necessarily create the infrastructure to do remarkable scientific research.

In this kind of scenario, a single human can be an astonishing machine of discovery and cognitive flexibility that is far more advanced than robotic avatars are at this point. We saw the adaptable human capability of the Apollo astronauts on the Moon. Even with only basic training in geology, the astronauts would notice anomalies and make connections between things in the lunar environment that would prompt vital discoveries, and they'd do it in seconds and minutes, not in hours or days or months. At our present level of technological know-how, a human is still the better choice for science. If we plopped a single scientist on Mars with a suite of scientific instruments and some ground transport, there's a pretty good chance that they'd resolve many of the outstanding questions we have about Mars's history and its indigenous life within a couple of weeks.

The other two schools of thought involve either placing lots of humans on Mars as permanent residents or avoiding the whole thing altogether. This is a fundamental divide in the rarified community of people who think about humanity's plans for Mars. On one side are those who feel that our overriding priority for Mars is to put humans on its surface and to establish a permanent human presence. Why we should do this, according to the advocates, is mostly for self-preservation because living on a planet like the Earth is inherently dangerous in the long term. Earth's history of devastating mass extinctions and uncontrollable hazards provides overwhelming evidence for that.

We know, from the statistics of asteroids and comets, and from the craters they cause on objects in the solar system, that the Earth will continue to be hit by objects large enough to wipe out much of life.[12] When the last major asteroid impact happened, sixty-five million years ago, it helped shuffle the large nonavian dinosaurs off the tree of life. Statistically speaking, we're somewhat overdue for another. We also know that Earth's own geophysical tumult isn't ending anytime soon, with super-volcanoes and other civilization-changing eruptive phenomena still possible. The coronavirus pandemic, which started at the end of 2019, also revealed incredible fragility in our global-supply-chain structure and our economic systems. And the more reliant we become on our technology and its ever-flowing information, the more vulnerable we are to events that destroy or debilitate that technology, whether those events are solar flares or self-made apocalypses from power shortages or from digital virulence.

The bottom line is that bad things are guaranteed to happen to our rocky planet and to happen again and again, and that the more complex and interdependent a species is, the more vulnerable it probably is. In that context, it doesn't seem quite so crazy to build some kind of fail-safe community in the wilderness regions of our solar system: a backup for our genes, our knowledge, and our brief lives. To boot, putting humans on Mars to stay there forever may actually be less costly than trying to bring them back again. For the same price, we get to expand the territory available to us and to create extraordinary opportunities for people on a whole other world that is otherwise just orbiting the Sun uselessly. However, there is a caveat that I'll return to later, which is that Mars faces its own catastrophes in events like massive asteroid impacts and may be even more prone to such things, so any backup is about playing the odds.

Even more ambitious ideas in this camp include the possibility of planet-scale geoengineering, or "terraforming," to turn Mars into a more hospitable place.[13] On the back of napkins stained by the spills of cheap cocktails, it is indeed possible for over-reaching scientists to show that devices such as giant orbital mirrors could be used to reflect sunlight to heat Mars's polar regions and unlock its frozen water and carbon dioxide. At a certain point, as the planet's atmosphere thickens with the input of these gases, the process begins to be self-sustaining. Temperatures rise because of an increased greenhouse effect, and raised atmospheric pressure allows for exposed water to remain liquid on the surface. Alternatively, massive amounts of water, carbon dioxide, ammonia, and methane could be "imported" by diverting comets and asteroids to crash into Mars and release these volatiles. Another possibility would be to dig enormous, deep trenches to let more of Mars's interior heat reach the surface. Or more subtle geoengineering could paint the polar caps with light-absorbing compounds, raising temperatures. Organisms might even be genetically engineered to chew up the harsh Martian oxidants and release useful chemicals into the environment, and so on.

The central problem with these ideas, apart from their wholesale disregard for any delicate scientific secrets held in Mars's indigenous landscapes, is that they ignore the enormous complexity and nonlinear reactivity of a planet's environmental systems. That includes climate systems as well as subterranean processes and volcanic cycles. These notions also ignore the wholesale availability and recycling of biologically essential elements like nitrogen that happens on a world like the Earth and that is so vital for a functional biosphere. Making a thicker atmosphere and raising temperatures on Mars is at best half a fix, but it could have many, many utterly unforeseen consequences that actually subvert any hopes of creating a temperate and habitable world.

Furthermore, although Mars may not be completely geophysically inactive, it is almost certainly far less active than the Earth and very different. Yet we know that geophysical cycles are essential for maintaining long-term conditions on rocky worlds. Here on the Earth there is a feedback loop that is thought to stabilize our climate over hundreds of thousands of years. Atmospheric carbon dioxide produces slightly acidic precipitation that weathers silicate rocks and traps carbon before recycling it through subductive plate tectonics and an eventual volcanic release back to the atmosphere. The hotter the planet, the more carbon dioxide is removed from the air, whereas lower temperatures slow that process, allowing the greenhouse effect to rewarm things. Earth's magnetic field, produced by deep geophysics, also helps prevent our atmosphere from being sputtered away into space as the solar wind batters its outer zones. Mars has neither of these mechanisms, nor is it likely to acquire them. In other words, even if hit-and-miss terraforming kind of works, the results are unlikely to last forever.

All of which brings us to the third philosophical camp for Mars's future—where we simply don't send humans at all. For starters, going off this world to create a backup for humanity and other terrestrial life sounds like we'd be bailing out on dealing with the problems of existence on the Earth. And the Earth is the single most ideal environment for us to exist in, at least for now. In this camp of thought, any Mars settlement or even human exploration would be a misdirection of attention and resources at a critical juncture in our history.

Pouring effort into the project of relocating and sustaining even a few dozen humans on Mars, much less a million (a figure sometimes discussed in popular media), can seem like a weird choice when there are so many ways we could improve and reinforce our existence on the Earth. Apart from addressing climate change and species extinction, those improvements would include doing something to address the appalling imbalance between rich and poor—the lucky and unlucky—and the minds and creativity that we lose because they never get a chance. Although creating a large human settlement on Mars might cost "only" a few hundred billion dollars (a figure that is a relative pittance compared to humanity's annual global military expenditure of more than $2 trillion), that's the kind of money that could transform the poorest regions on Earth. And if we are really determined to deal with very long-term survival, it is surely easier to build underground self-sustaining habitats on our planet of origin rather than out there in the interplanetary wilds. In this context, "Earth First" should be our new slogan.

Earth-firsters could also argue that selectively populating Mars in aid of existential backup is not just an unproven strategy; it also reeks of an antiquated and largely discredited worldview that echoes the flaws of imperial, colonial powers. This concern is amplified when ultra-rich and already controversial individuals like Elon Musk appear to put their stamp of approval on the idea of wholesale "pay as you go" human occupation, a concept perilously close to being indentured servitude on the off-world colonies.[14] And for many scientists, the notion of bespoiling a pristine world like Mars—which was once more temperate and conceivably life-bearing—before we've really studied it is extremely concerning, as is the idea of placing humans in obviously challenging circumstances when our capacity to explore the solar system using robotic devices is improving all the time for a fraction of the cost and risk.

This cost issue is particularly acute when you start to crunch the numbers for what it takes to carry people to Mars and get them back to the Earth again (assuming that people have a "right to return" from Mars). Fuel alone is a massive headache. Do you carry all the necessary combustible propellants like oxygen and methane the entire way down to Mars's surface, or do you find ways to refuel with supply drops or in situ manufacture from Martian resources?

Basic calculations suggest that for a two-person crew in a moderate-size capsule with life support, it would take more than 30 tons of propellants to hoist them back off the surface of Mars to a reasonably high orbital rendezvous point. From there, another craft would then have to take them back to the Earth. This is, in the vernacular of rocketry, a "high gear ratio" operation—with every kilogram of launched mass requiring several kilograms of propellant. It's worth recalling that when Werner von Braun described his "Mars Project" in 1952, he envisioned making 950 single-stage launches back at the Earth just to get everything needed into space before assembling it all and heading for Mars. Admittedly, that was for a fleet of ten spacecraft (and seventy humans) to go and return. But these numbers weren't just plucked out of thin air. Von Braun actually based these estimates of the scale of a bona fide Mars exploration on a real-world Antarctic exploration program that established a temporary US base on the Ross ice shelf in 1947.[15] That terrestrial effort involved 4,700 personnel, 70 ships, and 33 aircraft just to sustain humans in a polar environment (and still resulted in the accidental deaths of four people in the process).

There are also costs for the most basic of resources and support systems, with critical items being oxygen, water, energy, and food—followed by radiation protection, low-gravity mitigation, and mental health. These are all key commodities for humans on Mars. Take, for instance, the example of the oxygen that we, and all the food we grow, will need. On Mars, oxygen has to be generated either by stripping it out of atmospheric carbon dioxide or from splitting water molecules, and perhaps even by cooking up the Martian soil to release volatile compounds that can be further processed. On the currently active Mars rover Perseverance, an experiment called MOXIE sucks in air, compresses it, heats it, and breaks down carbon dioxide molecules with a special ceramic catalyst to generate molecular oxygen.[16] Based on this prototype, it's estimated that a bigger apparatus, powered by 25 to 30 kilowatts, could churn out 2 kilograms of oxygen per hour. Because humans need about 0.84 kilos of oxygen per day to survive, this system would support more than fifty people, or at least it would if they're not exerting themselves too much.

But a system like this demands its own infrastructure. Tens of kilowatts of power is a lot, and that means setting up something like a small nuclear-power station on Mars or deploying a few hundred square meters of solar panels. Breathing just pure oxygen is not a particularly good option either. Because of the risks of combustion in the environment and what our bodies are capable of handling, we need buffer gases like nitrogen, and we need to scrub our exhaled carbon dioxide from the atmosphere. The production of a breathable atmosphere is such a critical system that it would definitely need a backup—either a copy of the same machinery or a smaller machine for short-term use, as well as storage of enough oxygen to last people in the case of a truly catastrophic problem. But estimating the associated risk-benefit-cost balance of building these pieces is difficult, and the fact that there is no simple rescue from across interplanetary space adds to the problem.

Humans, like any biological life, also imprint themselves on their surroundings by generating waste. As a technological species, we've amplified this imprint by orders of magnitude, and humans are fantastic garbage generators. I looked into this in 2022, at which time we generated 1.2 billion metric tons of solid waste per year on the Earth, about half of which was organic waste from food and horticulture, and a third was paper and plastics.[17] Wastewater volume, from all sources, also totals a staggering 1,500 cubic kilometers per day across the planet.

You might imagine that we'd be doing a better job with waste in space, given the extreme burdens of hauling mass with rockets. But we've already dumped an estimated 180 metric tons of spent spacecraft and other refuse on the Moon—including twelve pairs of Apollo astronaut boots and an impressive collection of ninety-six plastic bags of poop, urine, and vomit. The International Space Station now does a good job at recycling about 93 percent of its onboard water, but the human occupants still generate about 0.63 tons of dry solid waste in packaging and other materials per year spent in orbit. That's only a little better than the 0.72 tons that the average American produces in a year, which is already much worse than in many other countries. On a space station, though, you have the option of jettisoning your garbage or packing it into capsules that return to Earth or are incinerated during reentry. You can also get resupplied relatively easily. On Mars, however, garbage and recycling are real conundrums, especially if a human presence is large. Materials and compounds will be precious, but unless we radically rethink how we use them and recycle them, we'll be in trouble.

What is the right path to take? All three approaches to Mars seem problematic for what they leave out or for what they need to succeed. I'm going to argue that there *is* an alternative, a fourth way of looking at things that balances actions and expectations somewhat differently. This path begins with the skills and tools we have generated over the past several centuries to examine and understand the Earth, and the incontrovertible fact that Mars is already a planet of machines and a planet of extremes.

These extremes are a result of Mars's distance from Earth and its native environment. These properties determine the fundamental nature of Mars exploration and raise possibilities not presented to any modern humans since the earliest migrations of our ancestors more than a hundred thousand years ago. Those ancient movements took *Homo sapiens* out of Africa, into the Middle East and Europe, and out into the Pacific wilds. We've reconstructed the pattern of these remarkable journeys using paleontological and genomic data to track the complex geographical ancestry of groups of peoples, which suggests that the last major exploration frontier for modern humans on the Earth was more than twenty thousand years ago.[18]

That frontier seems to have been in the northern and southern American continents, around the time of the Last Glacial Maximum, during a period when ocean levels were on average 400 feet lower than they have been since, unlocking land bridges and access points that are now hidden from view. Humans were able to reach continental landmasses that had never had such a species on them before, creating an astonishing opportunity for our ancestors but also presenting a completely unknown and fraught environment for survival. The arrival of humans also created new branching futures for the life that already existed in these places, reconfiguring megafauna and flora as people hunted for meat and modified the landscape.

In much more recent history, there have been modest human expansions into scarcely occupied regions of the planet. Perhaps the most unique frontier has been the enormous continent of Antarctica. This 14-million-square-kilometer polar desert, much of which sits at an altitude of more than 2 kilometers, is one of the least habitable parts of the Earth and with the least prior human presence. Today, the human population of Antarctica, consisting mostly of researchers and support personnel, maxes out at around four to five thousand people in the southern summer and drops to a dedicated population of some one thousand during the winter.

In this and in northerly polar zones, both long-term indigenous settlement and recent nonindigenous human settlement have confirmed that basic resources become an overriding central focus of survival. Modern polar exploration and short-term habitation have also played a key role in the emergence of wholesale conservation efforts and of a deeper scientific understanding of the systems that support Earth's climate and biosphere. In these extreme environments we are confronted with the raw, fundamental needs of human existence and how those needs affect our surroundings, whether in waste disposal or the inadvertent side effects on local ecosystems that come from building and occupying even modest amounts of infrastructure. All of this has shone an intense spotlight on the acute sensitivity of a place like Antarctica to local and global changes in temperature, as well as atmospheric and oceanic compositions. Life on the edge is where there is the most to learn.

That edge appears even in the peripheral zones of the southern polar environment, as the crew of the *Beagle* discovered in the nineteenth century. One of the most vivid episodes recorded by Charles Darwin was an 1832 visit to Tierra del Fuego, at the very southern tip of South America and separated by a thousand kilometers of ocean from the Antarctic Peninsula. Part of the purpose of that landing was to drop off three Fuegians who had, in effect, been abducted by Captain Fitzroy on the first voyage of the *Beagle* to this area two years earlier and were now being returned as "civilized" missionaries. That someone as sophisticated as Fitzroy would participate in this astonishingly cruel and inhumane practice was simply a reflection that it was not so unusual in the 1800s.

Despite Darwin's capacity for a deep and thoughtful kind of humanity, his outlook was also a product of the time and of the British Empire.[19] To read his accounts of reaching this southernmost archipelago of South America is to be reminded of that context and of the appalling racism that could be instilled even in someone of Darwin's sensitivities. His diary accounts of some of the "savages" and the "violent" culture that included "cannibals" conclude with thoughts like "Viewing such men, one can hardly make oneself believe that they are fellow-creatures, and inhabitants of the same world." What shocked Darwin the most was the seemingly desperate state of members of some of the Fuegian communities, who were barely clothed and fed in an environment that was wet, cold, and bereft of many obvious natural resources. Surely this was the last gasp of primitive human inhabitation?

Not surprisingly, we now understand the history and context of the people living in this region far better than anyone in Darwin's era did. Several distinct Fuegian groups occupied the archipelago: the Ona (Selk'nam), Haush (Manek'enk), Yaghan (Yámana), and Alacaluf (Kawésqar). All had complex histories going back at least several thousand years before the *Beagle* arrived, hardly the stuff of failure. But there is one aspect of Darwin's observations that was not tarnished by prejudice; life in Tierra del Fuego was probably never easy.

It seems that the early Fuegians were largely nomadic, with tribal groups having to roam as hunter-gatherers because of the limited flora and fauna, and because of the dramatic seasonal shifts in this windy subpolar oceanic climate. Across Tierra del Fuego, the highest temperatures during the year lurk around an average of 9 degrees Celsius (48° F), and cloudy weather leads to an average maximum of about six hours of daily sun at its height in the southern summer. Club Med it is not. Within the span of the archipelago, the land ranges from the spectacular tail-end peaks of the Andes, reaching as high as 7,000 feet, to swaths of glacial terrain with lakes and moraines, together with areas of tundra and some of the beech forests seen throughout Patagonia.

Tierra del Fuego is beautiful but extremely harsh. When Darwin saw what he considered to be near-naked, "stunted, miserable wretches" in the western part of this region, he was actually observing incredible human survival and the extreme costs of that survival. He recounts how "if a seal is killed, or the floating carcass of a putrid whale is discovered, it is a feast; and such miserable food is assisted by a few tasteless berries and fungi." This doesn't sound wonderful, but even though Darwin considered this as failure on the part of the Fuegians, it's now quite clear that they were finding a way to exist despite appalling odds.

There are lessons for Mars exploration in all of these things. In many ways Mars is akin to the Tierra del Fuego of ten thousand years ago, when the first humans ventured onto its islands. Consequently, Mars is a place where survival needs are going to take precedence over all other things unless something is done to create a hugely supportive infrastructure of tools and knowledge. Without that, any form of long-term human settlement on another world would be enormously unpleasant, and that would be the default mode of existence for a lot of the time.

Today, though, we know that one extraordinarily useful thing that we can do is to create the means to obtain, share, and build information about the world around us. That "knowledge infrastructure" is something that we know exactly how to implement through a planetary web of communications, remote sensing, and GPS navigational aids just like we have on the Earth. This is often overlooked in the discussion of Mars exploration, but one of the greatest necessities for a human and robotic presence is day-to-day knowledge of the planetary surface and its conditions, as well as a means to pinpoint locations and distances. As a thought experiment, imagine that the early Fuegians had cell phones, satellite imagery, and GPS: their opportunities to live in such a challenging environment would have been transformed. Instead of hit-and-miss nomadic movements based on past experiences, they could have made careful strategic choices in their migrations and managed their resources in ways that were grounded in measurements of geography, weather, and flora and fauna. Imperial powers, like the British, understood the enormous value of geographical and environmental data, and this was a primary motivation for the *Beagle*'s journey. On Mars we do now have the very beginnings of orbital monitoring, with some 99.5 percent of its surface mapped to a resolution of around 5 meters per pixel, and we have active spacecraft that can snap pictures fairly regularly. But what could come next are sensing systems like we have on Earth that allow us to track short timescale changes and seasonal variations across the entire planet and to be able to follow our own changing footprint as machines and habitats pop up across the surface. There are certainly challenges to this approach because, like the Earth, Mars is not a perfectly spherical body, so its gravitational field for low-altitude orbits creates issues for satellite stability. But from our experiences around the Earth and the Moon, we already know how we can adapt and choose among the best options.

Thousands of years of human thought on motion, energy, and gravity have led to us being able to capture extraordinarily detailed information about the Earth and all the ways that life occupies its niches, from the microscopic to the global. Mars is a world where we can *start* with that information, if we're smart about it. In doing so, we'd discover how that information changes the opportunities and chances for our species before we dismiss any option or plunge carelessly into something we'll regret. In this context, by attempting to live on Mars in the best possible way, we can force ourselves to undertake the most focused effort in our history to truly understand the nature of sustainable planetary habitation. Or to put this another way: it's human nature to not recognize what we need until we don't have it. Mars is a prime example of not having what we need; it is perhaps the most novel and extreme human-exploration challenge in recorded history—exceeding even the challenge of lunar exploration because of its remoteness from the Earth. That is exactly why we should learn what it takes to exist there.

This fourth philosophical approach has, I would argue, the greatest odds of success, it is the most natural extension of what we've already accomplished at Mars, and it can generate endless scientific discoveries. Beyond creating the infrastructure to map, monitor, and navigate Mars, this approach naturally calls for a blended human and machine presence: not a wholesale human settlement on Mars but a coexistence between the descendants of today's machine population and a tightly integrated, perhaps rotating population of humans and whatever other organisms it makes sense to have there. That would include plants and fungi, or colonies of single-celled organisms like algae to provide foodstuffs and medical resources. The human population would serve as scientists, explorers, observers, farmers, cultural ambassadors, and machine operators, as well as our backup against existential threats. They can be the conduits to help us find new insights about how to live on a planet that will be essential for anywhere that humans find themselves. Between a knowledge infrastructure and people and machines, we can make Mars the best-understood world next to the Earth.

Mars could also hold part of the key to teaching us what human-machine coexistence really looks like. This would continue a process that began with the development of microprocessors and software engineering during the lunar missions of the 1960s. Today, physical robotic augmentation on the Earth is arguably a fairly mixed bag. After all, it's not really critical to survival to have robotic drones that deliver your beer and chips to your front lawn. But on Mars, it might be. Machines can withstand the chemically caustic, high-radiation environment of the surface. They can be dispersed across many different localities yet act in concert by being networked together. Machines can—if properly developed and programmed—find and process resources, and distribute the products. Machine learning could help build the behavioral attributes for robotic devices to do these things efficiently and autonomously, turning them into properly native Martians. Human operators, or virtual copilots, might also provide robotic scientific probes and laboratories with the best of all worlds, adding cognitive flexibility to the machines' resilience to environments, allowing them to carry out precise but unpredictable tasks.

If we figure out how to keep terrestrial organisms alive on Mars for extended periods or even generations, it would also be ironically dumb to place all those living things in one place, where a single mishap or natural event could wipe everyone out. For example, Mars is at least as vulnerable to asteroid impacts as the Earth, and perhaps even more so. For that reason, any long-term survival or presence on Mars probably involves dispersal across its surface and a mobile robotic population to support the human population. This need, together with the extreme nature of living on Mars (at least as we currently envision it), may have startling consequences for how humans structure their systems of supply and economy, and of governance. Those lessons could become critical if conditions on the Earth deviate more and more from the conditions that produced our species in the first place, and they could provide essential information for what comes after Mars, as I'll describe later.

The long-term exploration of Mars may therefore be different than today's very clinical and limited version of space exploration. This fourth way is a layered approach that melds our talents at planetary observation, robotics, and computation with the messy and complicated nature of sustaining biology and handling the emergence of different social structures and opportunities. It involves making Mars equal to the Earth in terms of our mapping and sensing of an entire world. This pathway may not ever place millions of people on Mars, but it could serve all our needs: for backup, for the preservation of life on Earth, and for building scientific knowledge.

It's worth reflecting that fiction has cogitated on the perils and promise of Mars long before now. Take, for example, the highly refined and poetic literature of the writer Ray Bradbury. In 1950 Bradbury published *The Martian Chronicles* and created a Mars that held an ancient civilization in a state of exquisite decay that couldn't withstand the crude and voracious incomers from the Earth.[20] Bradbury used this setting to explore what is perhaps a uniquely American tale that charts the tragedy of colliding alien cultures that can barely comprehend each other. In his stories, the indigenous species of Mars are usurped into ghostly oblivion by the uncaring (or perhaps just unthinking) invaders from Earth, who are trying to escape their own tragic demise. Yet in the end, these interlopers themselves evolve into a new form of Martians. If you've not read any of these tales, you should; they're brilliant and perhaps prophetic. Although the real Mars doesn't seem to harbor complex life, it could still transform life from the Earth and undergo its own kind of transformation at the same time.

I'm going to end this chapter with a provocative statement that also helps explain why I think the "fourth option" described above is how we should be thinking about Mars. Mars may have a lot of our attention at the moment, but Mars could easily turn out to be a siren: a distraction calling for interplanetary sailors to scupper their ships on its cold, dry rocks rather than find their dream lives. The parochial European explorers of the nineteenth century often looked to sculpt foreign lands into simulacra of where they had come from, only to discover that different climates and geographies simply don't do that willingly. Whereas Darwin's expression of desire to witness the scenery of another planet, quoted at the start of this chapter, might finally be satisfied on Mars, we may also find that the real key to existing beyond the Earth looks altogether different and that Mars is merely a useful but temporary way station for developing the tools to do this.

The next step in that story takes us to the outer planets because it's in their realm that we've discovered how limited and parochial even our most vivid imaginings about the architectures and environments of other worlds have been. From the diversity of worlds to the nature of physical scale itself, there is very little that can prepare minds from a modest rocky planet for the true scope of the solar system.

—7—

THE MONUMENTS
AT THE EDGE OF
THE BELIEVABLE

One is astonished at the amount of creative force, if such an expression may be used, displayed on these small, barren, and rocky islands; and still more so, at its diverse yet analogous action on points so near each other. I have said that the Galapagos Archipelago might be called a satellite attached to America, but it should rather be called a group of satellites, physically similar, organically distinct, yet intimately related to each other, and all related in a marked, though much lesser degree, to the great American continent.

—Charles Darwin,
The Voyage of the Beagle

Imagine that you've been asked to describe the scale of human occupation of the natural world. We look around and think of all of the other eight billion humans milling around our towns and cities, against the background hum of vehicles and technology, in a vast planet-wide sprawl. So much of the Earth's raw material appears to have been dug and processed into bricks and metal that hold up skyscrapers and bridges, or refined into chemical compounds and extruded into billions of plastic forms.[1] Unless you're standing in the middle of a great desert or bobbing along in a small boat on a large ocean, it can seem like the human stain is taking up space everywhere.

But expand your horizons just a little way into the cosmos, and things change. You'll discover that all of the humans, soil, rocks, and oceans of Earth's outer crustal skin don't really amount to much. Not only are all of these items little more than a slapdash patina, amounting to a mere 0.5 percent of the mass of the planet, but the entire Earth represents only 0.0003 percent of the total mass of *all matter* in the solar system. In fact, all the planets, moons, asteroids, and cometary nodules are crumbs that add up to just 0.2 percent of everything. Most of our solar system's matter is instead contained in the Sun, whose mass is two thousand trillion trillion metric tons of fuming, primordial elements—only slightly tainted by the atomic products of prior stellar generations. In other words, not only has the entirety of life's four-billion-year history and humanity's two-hundred-thousand-year history played out in what amounts to the mineral scum coating a rocky planet like oil on a puddle, but all of this represents less than a billionth of what the solar system is made of.

The matter constituting planetary objects is also arrayed across distances that challenge our day-to-day conceptions of causality and time. Millions and billions of kilometers separate worlds and bodies, so the finite speed of light—a brisk 300,000 kilometers per second—becomes more and more limiting. When the Apollo astronauts bobbed around on the lunar surface, it took 1.3 seconds for their transmissions to reach us. The light of the Sun on the Earth has always taken 8.3 minutes to reach us. When our robots roam the surface of Mars, we cannot easily operate them in real time because signals take between 3 and 22 minutes to traverse space, depending on where Mars and Earth are in their orbits. And the most distant probe we have, Voyager 2, is now far enough away that a one-way-signal trip time is more than 22 hours. This is a form of causal isolation more profound and unsurmountable than anything you or I have ever experienced.

Even Charles Darwin might have been forced to rethink the "grandeur" that he saw in terrestrial life if he'd been confronted with the cosmic demographics of composition and scale. But there are places that can help us bridge the cognitive gap between life's present occupation of the solar system and these greater realities: worlds that restore some pomp and circumstance to the planetary leftovers from the Sun's formation, carry their own astonishing archipelagos, and offer the same kind of diversity of stories that Darwin saw in the Galapagos Islands on Earth. These worlds are another type of great continent in the form of the gas giants Jupiter and Saturn.

Simply stating that a planet like Jupiter is 317 times more massive than the Earth doesn't quite convey its absolute dominance of the planets of the solar system. If you were a distant alien astronomer seeking evidence of planets around the Sun, you might use the technique of Doppler variations—a measurement of the wobbling motion of stars in space as their orbiting planets exert a gravitational pull on them. Do so, and you'd find that the Sun exhibits a gentle rhythmic motion. This is a result of the orbit of the star around a common center of mass, a balance point of sorts, between it and its planets. Because the Sun is the dominant mass, that balance point is not offset from its center by much, but it's enough to cause a measurable wobble. The largest variation of the Sun's motion will peak at twelve meters per second, with a repeating period of ten years, and is caused entirely by Jupiter. That may not sound impressive, but the Earth, which is five times closer to the Sun, induces a mere nine *centimeters* per second of cyclical motion onto the Sun's mass. As an extraterrestrial astronomer, you might find that Jupiter is the first and perhaps even the only planet you would ever detect around this fairly nondescript star.

Here on the Earth, you can actually feel Jupiter's presence if you design the right experiment. One evening in New York a few years ago, when Jupiter was closest to the Earth and high up near the sky's zenith, its bright dot caught my eye. Intrigued, as perhaps only scientists can be, I wondered what Jupiter's gravitational pull was doing to me.[2] It turned out not much, but that's because my *weight*—the product of my mass and gravitational acceleration—is modest. But if you consider something like the Empire State Building, the results are a little more relatable in everyday terms. From across 600 million kilometers, Jupiter's gravity causes as much as a 12-kilogram (26-pound) change to the Empire's weight (with the common use of kilogram as a proxy for weight rather than mass). Admittedly, this is still incredibly small compared to the tower's total 330 million kilograms (730 million pounds) of weight, but considering how far away Jupiter is, it is an astonishing effect.

Jupiter is a monster, something we'd imagine to be from the cosmic depths, but it's right here in our solar system. In the 1968 movie *2001: A Space Odyssey*, created by Stanley Kubrick and Arthur C. Clarke, the fictional astronaut Dr. Dave Bowman arrives at Jupiter and encounters a mysterious black monolith in space. But in this scene, it is Jupiter that takes on the most alien aspect: an ominous and brooding presence surrounded by its brightly shining moons, dwarfing everything else around it and sheathed in rich swirling clouds and exotic hues. What better place than here for an incomprehensible civilization to place its sentinels?

Jupiter's enormous mass, and to a lesser extent Saturn's, even plays a role in the behavior of the long-term climate states of the Earth and—as I have mentioned previously—Mars. This happens through the perturbative gravitational effects of these massive worlds on the orbits of other planets and their spinning axes, precisely the effects that Laplace and Somerville developed the mathematical tools to describe two centuries ago. These perturbations influence Martian and Terran ice ages by inducing variations in orbital shapes and spin-axis orientations that shift in and out of phase over tens to hundreds of thousands of years—like an overlay of notes played by sets of musical instruments. These changes in a planet's spatial configuration alter how and when the planet absorbs sunlight, with profound effects on climate.

In many ways, the fact that we call Jupiter a planet, as if it bears some resemblance to the Earth or any of the inner worlds, is merely a historical accident from the original meaning of the ancient Greek term *planētes asters* ("wandering stars"). Jupiter is something that exists within a cosmic spectrum of increasingly large, gravitationally formed objects. If it were just ten times more massive and consequently hotter in its core, it would more closely resemble the smallest stars that we know of that are capable of initiating sustained nuclear fusion in their interiors. As it is, Jupiter's overall composition of elements is much more like that of the Sun than it is like that of the Earth. By mass, Jupiter is mostly molecular hydrogen and helium, in proportions of 71 percent and 24 percent, respectively, although its outermost, colorful atmosphere is noticeably laced with other stuff. That includes swirling cloud layers of light-colored ammonia ice particles, along with upwelling sulfur and phosphorus compounds, with ammonium hydrosulfides a little lower down, and varied water ices and vapor well beneath that.

We think that deeper inside, Jupiter is a progressively denser and hotter fluid dominated by liquid metallic hydrogen, an exotic and extraordinary material created under immense pressure (a substance that a few bold scientists have tried to re-create in earthly laboratories). Metallic hydrogen, as its name implies, has properties such as electrical conductivity and perhaps even superconductivity, and it is a consequence of the quantum physics of electrons and protons. Down toward Jupiter's center, this unfamiliar substance may be increasingly blended with iron and silicates, and the environment is compressed by the immense weight of all the material above it to reach temperatures as high as 50,000 degrees Celsius.

If you ventured down to Jupiter's visible cloud tops and hovered there, you'd experience a gravitational acceleration (and weight) two and half times that on Earth's surface, despite being at a distance from Jupiter's center of 70,000 kilometers—almost twice what a geostationary orbital radius is from the Earth's center. If you dropped about 60 kilometers farther down, you'd start to encounter other clouds of different condensed compounds and experience an atmospheric pressure that has already climbed to more than five times that on Earth's surface. The next 25,000 kilometers down would be a sunless journey into a sea dominated by liquid hydrogen, with pressures reaching half a *million* times those on Earth. By the time you encountered the metallic hydrogen phase, the surrounding pressures would be millions of times higher than on the Earth's surface, and temperatures would have already reached thousands of degrees. If any part of you had succeeded at remaining whole until now, it would be quickly dissolved into its elements and dispersed into this terrifying abyss.

Outside of Jupiter is another difficult type of environment for explorers. A consequence of Jupiter's magnificent internal composition is that it generates a powerful magnetic field. If you imagine that Earth's magnetic field is like that of a bar magnet, then Jupiter's field is about twenty thousand times stronger. As a result, not only does Jupiter have a strong interaction with the wind of particles streaming outward from the Sun; it also acts as a potent accelerator of particles itself. This powerful magnetic field drives massive currents of charged species like electrons, protons, and heavier ions in the space around Jupiter. This has serious implications for any technology or biology in the vicinity, with radiation levels high enough to cause extensive damage. Even at a distance of half a million kilometers from Jupiter, radiation levels are enough to give a human a fatal dose in less than a day, damage sensitive electronics and materials, and even cause the buildup of huge electrostatic charges that can destroy electrical circuits.

That's a major consideration for exploration because Jupiter hosts an enormous collection of moons (or more scientifically precisely: "natural satellites"), all worthy of attention. Today, more than ninety of the larger ones have been identified and named, and we think that there are thousands of additional satellites smaller than a kilometer across. The orbital radii of these satellites are spread across more than 24 million kilometers, and the very outermost objects take more than two Earth years to orbit around Jupiter. Some of these outer moons cluster together in retrograde orbits, in the opposite direction to Jupiter's rotation, that are highly inclined, or tilted, relative to Jupiter's equator (collectively called "irregular" orbits). They may be the disrupted pieces of larger bodies that have been captured into orbit around the gas giant, almost like ghostly planetary moths attracted to Jupiter's gravitational flame.

The largest of Jupiter's moons were first fully documented around New Year's Day in 1610, when the Italian astronomer Galileo Galilei used his new telescope to peer at Jupiter. Galileo spotted three, and then four, fainter points of light that seemed to move back and forth around Jupiter's bright disk, and he eventually realized that these had to be objects that orbited Jupiter itself. Although it is possible that evidence of these moons had been seen before—perhaps even two thousand years earlier by Chinese astronomers—it would have been hard for even the keenest human eye to chart their movements until the augmentation of telescopes.

What Galileo saw were the moons Io, Europa, Ganymede, and Callisto, four bodies large enough to be worlds unto themselves. Ganymede has a larger diameter than the planet Mercury, and Callisto is almost as big. They circle Jupiter in periods that range from Io's speedy 1.8 Earth days to Callisto's nearly 17 days, and their distances from Jupiter span a range from 420,000 kilometers for Io to 1.9 million kilometers for Callisto. For comparison, the separation of the Earth and Moon ranges from 360,000 to 400,000 kilometers, which took the Apollo astronauts some three days to traverse. This distance would fit easily inside the orbit of the innermost Galilean moon, Io.

The inner threesome of Io, Europa, and Ganymede have orbital periods that follow an almost precise ratio of 1 to 2 to 4—so for every single orbit of Ganymede, then Europa orbits twice and Io orbits four times. This behavior is called a mean-motion resonance, or Laplace resonance, and it is—as with most things orbital—a consequence of gravity. Like the resonances of vibrating musical strings responding to the right stroke of a bow, this orbital resonance is caused by the frequency of the repeated pull of each of these moons on the others as they orbit around. That rhythmic pull, in effect, locks in the pattern. Orbital simulations suggest that the present configuration may last another 1.5 billion years, until tiny deviations pull the moons into a new resonance—this time among the exterior triplet of Europa, Ganymede, and Callisto.

But this orbital dance is not the most interesting thing about these Galilean moons. Until 1973, when the Pioneer 10 spacecraft flew by Jupiter, we had only a superficial understanding of what these natural satellites were made of or the characteristics they exhibited.[3] By the time that Voyagers 1 and 2 passed through the Jovian system in 1979, it started to become very clear just how special these places were. With eight color filters, the scanning Voyager cameras and sensitive infrared detectors provided the first close-up views of the Galilean satellites and their compositions, and those views were quite something.

In a routine task of processing a Voyager 1 image used for navigation, astronomer Linda Morabito, working at JPL, realized that the image had serendipitously captured an enormous plume of erupted material coming from the surface of Io.[4] Previous theoretical work had hinted at the possibility that Io might be geophysically active. Its surface seemed to be "young" in the sense that there were few craters accumulated from infalling asteroids, suggesting that the surface was being restored from within. But the discovery of active volcanism shooting material into space took things to a whole other level.

Io is a seething volcanic world, with the most geological activity of any planet or moon in the solar system—harboring an estimated four hundred active volcanoes across its sulfur-rich, yellowed surface, with more than a hundred mountain peaks that are taller than Mt. Everest. Yet Io is small enough in planetary terms that it should be a cold inert lump. (It's about 29 percent of the radius of the Earth, thus similar in size to our moon.) Instead, as a consequence of the Laplace resonance in the Galilean moons' motions, Io's orbit is maintained in a more elliptical shape than it would be otherwise, and it also spins so that its same face always points toward Jupiter. As a result, tremendous tidal stresses act on this moon during each orbit, forcing bulge-like distortions that lift and drop the surface by a hundred meters. The friction of that mechanical kneading creates heat throughout Io's interior. In other words, Io's volcanism is quite literally powered by the flow of momentum and energy from orbital motions into its rocky body.

A consequence of Io's volcanism is that plumes of sulfur-rich material are ejected high enough that their constituent atoms get caught up in Jupiter's intense magnetic fields—the planet's magnetosphere. This constantly feeds a giant, torus-like orbital cloud of sulfur ions around Jupiter, and some of those particles end up transferring onto the surfaces of the other moons, like Europa. That filthy spray gives Europa a dingy beige-to-red coloration in places. That coloration would otherwise be highly unexpected because, utterly unlike Io, Europa's entire surface is made of frozen water.

Prior to the Pioneer and Voyager flybys, Europa, as a moon slightly smaller than Io, was also little understood. In the 1950s the famous Dutch astronomer Gerard Kuiper and others used pioneering infrared-sensitive telescopic instruments on the Earth to determine that both Europa and Ganymede were coated in a large amount of water ice.[5] But it was the Voyager images, particularly from Voyager 2 as it passed by Europa at a distance of 200,000 kilometers, that proved transformative for our understanding of this moon. Europa was indeed all blushed water ice and was incredibly smooth compared to other worlds. But that surface was ridden with dark streaks crossing the whole body. These streaks, or lineae, appear to be where the surface crust has split and shifted, and warmer ice has oozed upward. There are also large regions of chaotic, angular terrain where great kilometer-sized rafts of ice have fractured and moved around in a plain of jumbled ice blocks.

These observations by the Voyagers, along with measurements of how Jupiter's magnetic field behaves around Europa, made it clear that this was a world unlike any previously explored. Although we've merely flown by Europa, the tools of mathematical physics and our knowledge of geophysics and oceanography have allowed us to create confident models of its properties. Beneath the moon's frozen crust—which is perhaps 15 kilometers thick—is an inner liquid-water ocean that extends another 100 kilometers down: a dark abyss that contains about twice the volume of water that is in all of Earth's oceans. We now have data (some from decoding the chemistry of the blushed surface where Io's sulfurous filth rains down) to suggest that this ocean is saline, about as salty as Earth's oceans. That salt comes from where the bottom of the ocean meets and dissolves some of Europa's rocky bulk, composed of the same kind of minerals we see everywhere in the solar system.[6] Europa sustains this interior ocean because of tidally induced heating, a toned-down version of what's happening on Io, making this another world that is powered by gravity and motion.

Because of these discoveries, Io and Europa get a lot of attention, but the other Galilean moons, Ganymede and Callisto, are equally extraordinary. Ganymede is also an ice-caked world, larger than the planet Mercury and with its own deeply hidden interior ocean that may be the largest in the solar system. Its surface, like Europa's, is coated by effluvia from Io and is far darker on its trailing orbital side. With a robust, fixed magnetic field, Ganymede seems likely to have a hot, partially molten rocky core, and it can even generate glowing auroras above its surface as Jupiter's flowing radiation and magnetic fields interact with the moon's own magnetosphere.

Callisto is the third-largest moon in the solar system, after Ganymede and Saturn's Titan, and has the dubious status as one of the most heavily cratered objects yet known, with a deeply scarred and darkened icy crust. The largest crater on its surface is the imposing Valhalla impact basin, a 600-kilometer-wide central bright spot surrounded by concentric, ripple-like circular fractures reaching out some 1,800 kilometers from the center—the remains of a cataclysmic event on Callisto more than two billion years ago. Even though the accumulation of craters on its pockmarked exterior tells us that Callisto's modern surface is quite inert, it too may have a deep liquid-water ocean beneath a depth of 150 kilometers of ice.

As well as the Galilean moons, there are dozens of Jupiter's natural satellites that are worthy of attention, and perhaps all of them are. The next-largest after the Galilean moons is Amalthea.[7] This reddish potato-like object is about 200 kilometers long across its major axis and orbits much closer in than the Galilean moons, at just 200,000 kilometers from Jupiter. Dust ejected from Amalthea's dirty ice surface also makes a tenuous ring around Jupiter and may be a result of the moon's internal heating, either by gravitational tides or because it is sweeping through Jupiter's magnetic field like a piece of a giant electric motor. Finding an icy moon this close to Jupiter is a puzzle because four and a half billion years ago the young gas giant would have glowed like an 800-degree-Celsius bar heater, boiling away nearby water. It seems likely that Amalthea is an interloper, an object formed in colder parts of the solar system and then somehow captured deep inside Jupiter's gravitational well.

There are similar likely interlopers like Thebe or Metis. The latter is a 50-kilometer-wide reddish object that is the innermost of all of Jupiter's known moons, racing around at 31 kilometers per second about 60,000 kilometers above Jupiter's cloud tops. Toward the outer edges of the Jovian moon system is 140-kilometer Himalia, one of the irregular moons, which circles in an 11-million-kilometer orbit together with a group of at least eight other satellites.[8] This group, and Himalia itself, may be the broken remains of a larger asteroid that was snatched out of a sun-centric trajectory and captured into orbit around Jupiter. These moons, like the majority of the outermost ones, move around Jupiter in a retrograde sense. This feature is also what suggests that they weren't born where we find them but were captured.

Jupiter's heft makes it the gravitational gateway between inner and outer worlds. Its presence was capable of sculpting the very structure of our solar system as it formed billions of years ago. Its gravitational pull helps drive the orbital variations on Earth and Mars that take these worlds in and out of ice ages. It shapes the orbits of objects in the asteroid belt beyond Mars and can divert the trajectories of bodies like comets that fall inward to the Sun or are on their outward journey again.

Until we fully understand the influence of this giant planet, we won't be able to complete the evolutionary chart of our own cosmic origins. More than this, Jupiter, its giant companion Saturn—itself 95 times more massive than the Earth—and their joint complement of 240 named satellites offer us a glimpse into alternate cosmic environments just as relevant as our present ones. This mental shift takes some getting used to. Europa, we say, has a deep, dark, and salty ocean that might resemble the Earth's deepest ocean environments. Or we say that Titan has an atmosphere at its frigid surface similar in pressure to Earth's and might produce cold, complex organic chemistry like on a primordial Earth. But when we make these comparisons, we also miss something: the opportunity to rethink everything we know about our place in the cosmos, to shake things up in the most Darwinian way possible.

For example, scientists and laypeople alike will talk about the Earth as a lonely, watery oasis in space. Yet across the solar system the majority of liquid water actually exists in the interiors of places like Europa and Ganymede. If we add up all the locations where these inner oceans should be—a landscape that includes moons and even objects beyond Neptune's orbit—then approximately sixteen times the volume of Earth's liquid oceans exists elsewhere in the solar system. In the larger environment around the Sun, Earth's surface water is both a proverbial drop in an interplanetary ocean and an outlier because it's unenclosed, almost directly exposed to a ghastly universe!

AFTER VOYAGERS 1 AND 2 scooted through Jovian space in 1979, there have been only a few other flybys. These included the Ulysses solar probe in 1992, which used Jupiter for a gravity-assist maneuver to reorient its orbit out of the plane of the solar system in order to fly above the Sun's polar regions. But in December 1995, the United States' Galileo probe became the first spacecraft to ever enter into orbit around Jupiter, on a mission that would last for eight years.[9]

Galileo was a 2-ton, 6-meter-long juggernaut bristling with scientific instruments. It was also a direct descendant of earlier generations of probes, adapting and improving technologies that had flown on the Mariner and Pioneer missions. To stabilize itself in space (because these probes are prone to tumbling or rotating in undesirable ways in microgravity), one whole section of Galileo spun three times a minute, carrying its sensors and experiments around and around.

Somewhat unusually, the rest of the craft was kept in a fixed orientation so that the cameras and spectrometers could lock onto their targets.

All of this complex deep-space hardware was controlled using six radiation-hardened microprocessors (made with radiation-proofing technology that I'll talk more about later), with four of these housed in the spinning side of the craft and two on the other. These chips were configured to provide redundancy for handling critical data and spacecraft commands in case any one piece failed. In the end, though, the biggest issue faced by Galileo wasn't this kind of electronic failure; it was in the mechanical unfurling of its umbrella-like high-gain communications antenna. Part of that story lies in the fact that in order to get to Jupiter, Galileo had to make a series of gravity assists using Venus and Earth rather than take a faster, more direct route. This was a consequence of snowballing delays brought about by the Challenger shuttle disaster in 1986 and the limited Delta-v budget of the spacecraft's propulsion system. Because of these maneuvers, the first time the command to unfurl the antenna could be sent was nearly two years into the mission, when Galileo was finally headed directly to Jupiter. But when that process was triggered, some of the ribs of the umbrella stuck, meaning it could open only halfway.

The loss of this antenna was a big deal because without it, the default mode of data transmission from Galileo back to the Earth was ten times slower—a painful bottleneck. Eventually, the project engineers traced the cause of the failure to decade-old lubricants in the mechanism made from dry molybdenum disulfide. These lubricants and the parts they coated had suffered unanticipated degradation. In order to deal with mission delays and to save money, Galileo had been shipped three times on a flatbed truck across the United States, back and forth from California to Florida, subjecting everything to vibrations and erosion that nobody had planned for. It was a case study in the unintended consequences of even the most innocuous-sounding decisions made in developing such complicated devices as robotic space probes.

But what this failure did was stimulate an aggressive effort to find a work-around. The Earth-based receiving stations were physically upgraded to handle weaker signals, and engineers quickly developed far more sophisticated ways to compress the information electronically before it was transmitted—using Galileo's onboard mechanical four-track tape recorder. The results were a spectacular success, with an effective data rate nearly ten times *higher* than the original specifications for the mission had it worked flawlessly.

Galileo also carried a separate 340-kilogram battery-powered atmospheric entry probe that was released months before arrival, when the mission was still 50 million kilometers away from Jupiter. This was part of an audacious choreographic plan for Galileo's eventual encounter with Jupiter. Months later, and just four hours before the entry probe would hit Jupiter's atmosphere, the main Galileo spacecraft used a gravity assist from Io to help get it ready to orbit around Jupiter and get properly positioned to receive the atmospheric probe data. The moment these data were fully captured, the spacecraft had to fire its engine for forty-nine minutes to make the final Delta-v adjustments to insert itself into orbit around Jupiter.

When the entry probe encountered Jupiter's upper atmosphere, it was traveling at 48 kilometers per second (a speed that would take you from New York to Los Angeles in 90 seconds). A specially designed heat shield helped it decelerate to a subsonic speed in just a few minutes, until a parachute deployed to slow the probe down to 430 kilometers per hour. This was a hair-raising maneuver because the probe's entry velocity was about four times that of the Apollo command modules as they returned to the Earth, and also because of Jupiter's powerful gravitational acceleration.

An hour later, at 180 kilometers beneath the cloud tops, the probe's signal was lost. With no solid surface, the entry probe would have continued to drop until the surrounding pressure reached around 5,000 times that at Earth's surface and the temperature reached over 1,700° Celsius, crushing and melting the device. Before its end, though, the probe returned a wealth of scientific data and surprises. These upper Jovian atmospheric layers were hot and dense, and they moved with wind speeds as high as 530 kilometers per hour. There was less helium gas and a lot less water vapor than expected—and because water vapor plays a large role in the temperature and structure of a planet's atmosphere as it condenses or evaporates, this was a major puzzle. Later measurements have suggested that water vapor is much patchier and less uniformly distributed than expected in Jupiter's upper atmosphere, so the Galileo entry probe may have just plunged down a dry zone by chance.

Galileo's orbital entry around Jupiter was also meticulously planned because the spacecraft had to pioneer a strategy for how to orbit a giant planet surrounded by an intense and destructive radiation field. The approach was simple in principle: by maintaining a highly elliptical orbit, it was possible to swoop in to be close to Jupiter or its moons. At this periapsis, or near point, Galileo would be moving fast and wouldn't linger in the most intense parts of the radiation fields. When it looped back out to the lower radiation environment of its slow orbital far point, the probe could make a leisurely transmission of data back to Earth. In its nearly eight years at Jupiter, Galileo made a total of thirty-four orbits, most of which swept the spacecraft across distances of at least twenty times the separation of the Earth and the Moon. The closest approaches to Jupiter, just within the orbit of Io, made the radiation dangers very apparent, with Galileo experiencing a number of temporary glitches brought on by radiation flipping bits in Galileo's electronic memory and interrupting processor functions.

By the time that Galileo was decommissioned by flying into Jupiter's atmosphere in September of 2003—to prevent it from ever inadvertently crashing on and contaminating the icy moons—it was a radiation-riddled shell of its former self, with various systems suffering from malfunctions. But this way of exploring Jupiter's harsh environment had proven extremely successful, and in 2011 the United States launched the Juno mission, a 3.5-ton spacecraft that was the first solar-powered spacecraft to the outer planets. This was enabled by advances in solar-cell technology, but even so, Juno had to be built with three 9-meter-long solar arrays covering a 50-square-meter area just to generate a modest 480 watts of electrical power from sunlight that is less than 4 percent as bright as it is at the Earth. Juno also faced its share of deep-space, long-duration exploration challenges. When the spacecraft arrived at Jupiter in 2016, the plan was to use its propulsion system to reduce its large elliptical capture orbit around the planet to a much tighter one. But problems with slow-opening valves in the propellant pressurization system made that too risky a maneuver, and instead the spacecraft has gradually modified its original fifty-three-day orbit around Jupiter using flybys of the moons to reduce this to thirty-two days, swooping around the planetary poles and overflying other areas just a few thousand kilometers above the atmosphere.

One of the most intriguing scientific results from Juno has been not about Jupiter's swirling exterior features but about the planet's deepest structures and formation history.[10] Because of Jupiter's major role in sculpting aspects of the solar system's orbital architecture (about which we'll talk more later) and its influence on the properties of planets and moons, what may seem like esoteric science is directly relevant to our assessment of life's expansion into space. Since the 1920s, researchers had relied on theoretical models of how the insides of giant planets might be layered. The first proposal for metallic hydrogen in these worlds was made in 1935, hot on the tail of revolutions in quantum mechanics and condensed-matter physics that could explain why such a substance might exist. It was justifiably assumed that the heavier elements of Jupiter's interior, like iron or rock, should—by virtue of their greater density—drift, or sediment, toward the core of the planet. That distinct core region would have a mass some ten times that of the entire Earth. But as more and more modern data on Jupiter have been gathered, the story has become more mysterious.

For example, helium may not be uniformly mixed in with the dominant component of hydrogen in Jupiter. The Galileo entry probe found less helium than anticipated, and there are reasons to think that helium actually "rains out" as condensed droplets that fall or drift inward, depleting the amount of helium at higher altitudes. With Juno we have made greatly improved measurements of the deep internal distribution of Jupiter's mass. These measurements use techniques similar to the ones used to examine Earth's gravitational field. Juno has a pair of finely tuned radio receivers and transmitters, operating at two microwave frequencies, that allow ground stations on the Earth to measure the subtle Doppler shifts in these frequencies as the spacecraft orbits. Those frequency shifts reveal how the orbit is distorted because Jupiter is neither a uniform solid sphere nor a spherically symmetrical object. From these data, the interior distribution of Jupiter's mass can be mathematically reconstructed. The result is that Jupiter's core appears to be strangely "fuzzy," a diffuse mush of material up to almost half the planet's radius. One implication is that Jupiter's original formation from protoplanetary material and its subsequent history are different from what had been widely assumed. It is possible that part of Jupiter's growth took place by gobbling up billions of small objects—like the smallest asteroids and pebble-sized material in the solar system today. These objects have, in effect, been dissolved as they sank toward the planet's center, creating a fuzzy mix of light and heavy elements. Another possibility is that a young Jupiter may have been hit by another massive body that burrowed down to the core and disrupted it, mixing those heavy elements into the surrounding hydrogen and helium, a bit like the way a pastry chef disrupts an egg yolk into the surrounding flour.

In 2023 the European Space Agency launched its Jupiter Icy Moons Explorer (JUICE) mission for an eight-year interplanetary haul to the Jovian system. As we've seen, to make the journey JUICE uses an elaborate, fuel-saving sequence of four planetary gravity assists from the Earth and Venus. This route is so elaborate that JUICE will actually pass through the asteroid belt between Mars and Jupiter twice before it finally intersects with Jupiter in 2031.

After three years in orbit around Jupiter, the spacecraft will then enter orbit around Ganymede for an anticipated year or more of up-close study before being deliberately crashed onto that moon's surface. To do all of this maneuvering, JUICE carries around three tons of propellant. In the meantime, in late 2024 the United States launched its Europa Clipper mission to arrive at Jupiter in 2030, just ahead of JUICE.[11] As its name implies, this mission's primary goal is to study the potential ocean world of Europa, deploying high-resolution cameras, ice-penetrating radar, and a host of other remote sensing techniques. Depending on whether the Juno mission is still operating in the 2030s, it is possible that Jupiter could host three robotic probes at this time.

But the realm of the gas giants isn't Jupiter's alone. Until 2004, Saturn had simply had spacecraft fly past it: Pioneer 11 in 1979 and Voyagers 1 and 2 in 1980 and 1981. That changed on July 1, 2004, when, after a seven-year cruise from Earth, the three-ton internation-ally designed and built Cassini probe inserted itself into orbit around Saturn for what would be an extraordinary thirteen years studying this world along with its record-busting complement of more than 145 charted moons, thousands of smaller natural satellites, and its famous and beautiful ring system.

Even more so than the missions to Jupiter, the Cassini project could be the best example yet of what it might be like to send a probe to another star altogether.[12] Such a mission would be a truly remote expedition, where—as with Cassini—robotic instruments survey and study an entire system of worlds, eking out more and more surprising discoveries and pushing technology to its limits. And just like Cassini, this future probe would almost certainly have to invent new ways to apply tools that were designed with very different expectations a long, long way away.

This trip to Saturn, known more precisely as Cassini-Huygens, began to take shape in the early 1980s as European and US scientists formulated plans to explore the outer planets in the immediate wake of Voyagers 1 and 2 flying by Saturn. By 1989, after a great deal of scientific and political wrangling, the European Space Agency, NASA, and the Italian Space Agency started to work in earnest on a mission. The design became a 7-meter-long and 4-meter-wide primary spacecraft, combined with a 1-meter-wide, 350-kilogram probe called Huygens, honoring the famous eighteenth-century Dutch astronomer who discovered the Saturnian moon Titan. Fittingly, the Huygens probe would enter Titan's thick atmosphere and descend to the surface.

Like the Galileo mission, Cassini-Huygens was a supremely complex project. The main spacecraft had more than 14 kilometers of wiring onboard to connect its control systems and a dozen scientific instruments and experiments. It had a powerful, doubly redundant rocket engine for planetary maneuvering, and it carried 33 kilograms of plutonium-238 oxide pellets, whose red-hot nuclear warmth drove a thermoelectric generator to produce as much as 700 watts of power. Launching this much plutonium fuel on a large rocket from the Earth was clearly a risky business, but at Saturn, nearly ten times farther from the Sun than the Earth is, sunlight is a hundred times less bright and solar power is largely impractical.

Dates and numbers are important for the Cassini mission because they write a story of what long-term space exploration looks like. The first construction efforts on the project had started in earnest in 1989. The mission launched in 1997 into a heliocentric orbit that took the probe by Venus twice in 1998 and 1999, then the Earth in 1999, and then a Jupiter flyby in 2000, using each of these gravity assists to reach Saturn in late June of 2004. Six months later, on Christmas Day, the Huygens probe was released into its own orbit around Saturn before intersecting Titan, where it entered the moon's atmosphere in mid-January of 2005. The main Cassini spacecraft then spent the following thirteen years in a sequence of shifting orbits around Saturn that took it past moons, past rings, and even up over the planetary poles in a voyage that became known as the ball of yarn because of the appearance of these pathways, as shown in Figure 7.1.

In total, Cassini made 294 orbits around the Saturnian system, with each taking between 6 and 30 days. The probe fired its engines 360 times and gathered 635 GB of data, making 162 flybys of different moons. Yet despite all of this, Cassini's 13 years at Saturn spanned only a third of Saturn's own 30-year orbit around the Sun.

As with the Galileo and Juno missions to the Jovian system, Cassini's time at Saturn has added a wealth of data and insight to gas-giant planets and their environments. At a third of the mass of Jupiter, Saturn is in many ways a much gentler world. With less circulating internal metallic hydrogen, Saturn has a twentieth of the magnetic

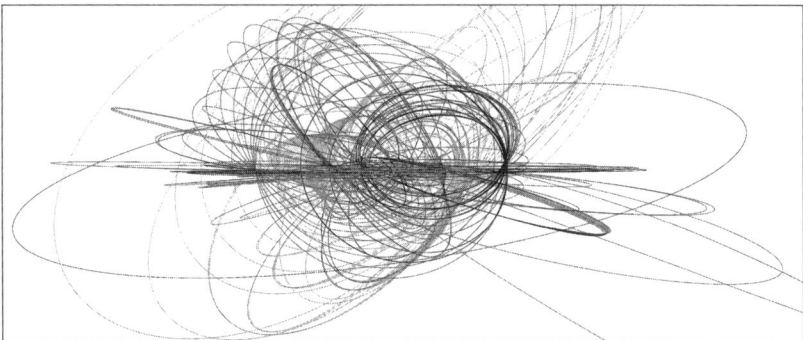

Figure 7.1. Illustration of the Cassini orbits (the "ball of yarn") during the probe's thirteen years at Saturn. Each orbit lasted between a week and several months. (Credit: NASA/Jet Propulsion Laboratory-Caltech, Erick Sturm.)

field that Jupiter does, and it consequently doesn't generate the same kind of fearsomely energetic particle radiation that bombards many of Jupiter's large moons. Although this is a massive world, it is also much "puffier" than Jupiter, with an overall density less than that of liquid water. That low density and a fast spin (a day is about ten hours) allow centrifugal forces to distort it from a spherical shape, so it bulges out toward its equator. Saturn's cloud tops of ammonia crystals appear far less varied and visually complex than Jupiter's, but it is still wrapped by bands of jet-stream–delineated atmosphere and sometimes has vast atmospheric storm systems like the Great White Spot, which appears as a discoloration roughly every Saturnian summer solstice and has been observed by astronomers since 1876.

Cassini's data on Saturn's majestic ring system, which comprises multiple annuli of tiny orbiting particles that are mostly composed of water ice, have provided a playground for scientists interested in the same kind of complex orbital dynamics that fascinated people like Laplace and Lagrange. The rings are far from perfect in their symmetry, and Cassini witnessed changes and deviations as rings and moons interacted, and even what may be ethereal electrostatic effects playing across the structures.

Eventually, as a mission finale, on September 15, 2017, the Cassini spacecraft was deliberately sent on a terminal dive into Saturn itself after orbiting twenty-three times through the inner gap between Saturn's rings and its upper atmosphere. As had been done with Galileo at Jupiter, destroying Cassini in the heat of atmospheric reentry was done to deliberately remove it from any potential future contaminating collision with Saturn's moons, with good reason: among Cassini's many, many scientific discoveries is the possibility that some of these worlds could actually harbor alien life.

A prime example is the moon Enceladus, which, similarly to Europa and other icy bodies, seems to consist of a crust of frozen water and an internal ocean of salty, liquid water.[13] In 2005 Cassini discovered plumes of icy particles erupting high above Enceladus's 500-kilometer-diameter surface. These "cryo-volcanoes," located on the moon's gnarled southern landscape, seem to be venting material from that inner ocean. In fact, the vented material doesn't disperse altogether into space; it persists in tiny frozen particles that form Saturn's E-ring. With a diameter of 300,000 kilometers, that ring may represent the largest persistent distribution of ocean water in the solar system.

When Cassini was redirected to fly directly through Enceladus's plumes, its instruments, though never designed for this purpose, were able to detect water vapor along with small amounts of nitrogen, carbon dioxide, and ammonia, as well as simple hydrocarbons like methane and acetylene. In the years after the mission, intense scrutiny of these and other data has revealed the presence of silicate particles and other compounds that appear to have been produced deep inside the moon, where its warm rocky core meets the ocean, akin to hydrothermal vent systems in Earth's own oceans. Not only does this suggest an environment warm enough for biochemistry; it also suggests that there are sources of chemical energy that enterprising organisms might feast on.

Titan, the target of the Huygens probe, also turned out to be a place of outstanding interest. Prior to the arrival of Cassini, Titan was known to be something special. Like the other major moons of the gas-giant planets, Titan is large, with a diameter greater than Mercury's and only 25 percent smaller than that of Mars. By the 1940s, astronomers like Gerard Kuiper had discovered that Titan was unique among all moons in the solar system for having a thick atmosphere, and that atmosphere appeared to contain methane. Later data, including from the flybys of the Pioneer 11 and Voyager probes, revealed that Titan's atmosphere is mostly nitrogen, methane, hydrogen, and a few other compounds, and it's thick enough that the surface pressure on Titan is actually 50 percent higher than the surface pressure on the Earth. But this far from the

Sun, there is little radiation to warm worlds, and Titan has a peculiar climate state that is both a greenhouse effect because of the methane in its atmosphere and an anti-greenhouse effect because of the high-altitude hazes of organic compounds that reflect more radiation than expected. Consequently, the surface of Titan is only a little warmer than if it had no atmosphere: a frigid −180° Celsius (−290° Fahrenheit).[14]

Yet Titan is far from being a static, frozen environment. Cassini and Huygens discovered that the moon has active rivers and lakes, rain and weather, and (thanks to an axial tilt and Saturn's orbit around the Sun) seasons of winter, spring, summer, and fall that each last around seven years. Those rivers, lakes, and rain are not formed by water, though. At these temperatures, water has the consistency of rock, but methane and ethane and other hydrocarbons can transition among liquid, vapor, and even solid forms. In other words, Titan is a world where these hydrocarbon compounds play the role that water does on the Earth, replacing a hydrological cycle of flowing, evaporating, and condensing and raining water with one of a compound like methane.

When Huygens plunged into Titan's atmosphere, it decelerated and deployed a large parachute, letting it drift down to the surface over a period of more than two hours. In this descent the probe was witness to an environment far more alien than in a world like Mars. As Huygens sampled and measured Titan's atmosphere, its rocking and spinning capsule also built up a digital library of snapshot images. Researchers stitched these together to create aerial views of the moon's deceptively Earth-like topography, as shown in figures 7.2 and 7.3.

The technical challenges of this descent and landing were immense, and they had their own share of the kinds of hair-raising problems that complex missions encounter. One problem came to light not long after Cassini-Huygens launched in 1997. A subtle but potentially catastrophic design flaw was discovered in the communications systems that would relay the probe's data (from its

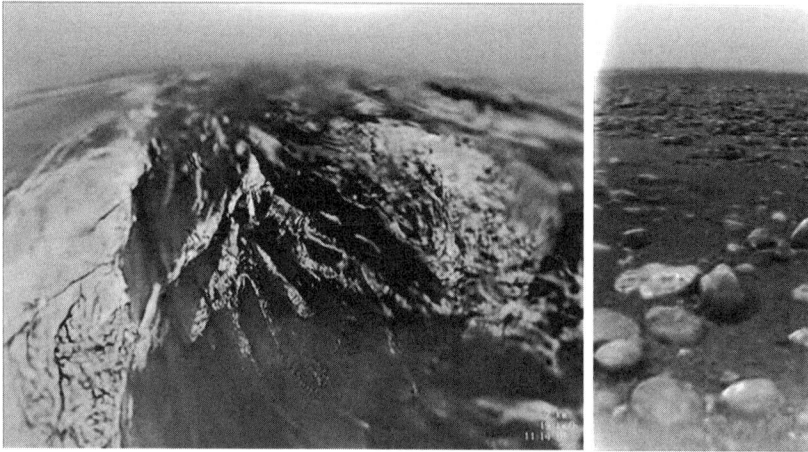

Figure 7.2. Aerial view from an altitude of 8 kilometers above the surface of Titan, created from images taken by the Huygens probe during its descent. The image is distorted (curved) because of the way it was assembled, but different types of surface are clearly visible, along with what may be river features on the lower left.

Figure 7.3. An image taken by the Huygens probe on the surface of Titan. The rocks are likely made of water ice and are around 10 centimeters across. The "soil" seems to be ice grains coated in darker hydrocarbon compounds. Thermometer readings on the probe suggest that its warmth was quickly lost to the surround- ings and that the ground was therefore damp with hydrocarbons. (Credit: ESA/ NASA/JPL/University of Arizona; surface image processed by Andrey Pivovarov.)

comparatively weak transmitter) to the main Cassini spacecraft and then onward to the Earth.

This flaw was positively baroque in its intricacy. The original plan for releasing Huygens down onto Titan involved a very spe- cific flight path that would cause the probe to accelerate away from the main Cassini spacecraft. Consequently, Huygens's radio trans- missions would be Doppler shifted—their frequencies changed by this relative motion—and the receiver on Cassini had been cleverly designed (or so everyone had believed) to accommodate this precise range of frequency shifts. But the built-in signal decoding system (the actual firmware onboard the spacecraft) had *not* been constructed to allow for the changing transmission frequency and for the fact that the data *rates* (the cadence of blip-like ones and zeros of binary data) would also change, which would confound the electronics and garble the data.

It's not hard to imagine the consternation among mission scientists when this flaw was discovered, rendering a $400 million probe unfit for its purpose. Luckily, with so many variables and factors in any space mission, there are often ways to rethink things, and in the case of Cassini-Huygens there was room for a fix. Even if the hardware and firmware of the two spacecraft couldn't be modified, their motion could be. Instead of dropping Huygens at the originally planned time and trajectory, it was detached from Cassini a month later, in December 2004, causing its motion relative to the main spacecraft to be nearly perpendicular and zeroing out almost all of the frequency and data-rate shifts from the Doppler effect, allowing the hardware to work just fine.

As Huygens floated down to Titan's surface, it saw a world with gentle mountains and shorelines abutting lakes and seas, and with plains of dark, fine, sand-like hydrocarbon soil that even forms dunes in places. Other regions appeared to be geologically young, perhaps molded by the cryo-volcanism of molten ammonia and water. When Huygens finally landed, it crunched into a crusty, sticky surface of finely grained ice colored by dark hydrocarbons and peppered with small rounded rocks of solid water. If you didn't know better, you'd think you were looking at the rocks and scenery that we find on Mars, or even Venus, but Titan really exists in an alternate dimension of temperature and chemistry.

Titan's cold environment is intriguing because complex chemical reactions and complex molecules can readily exist there. Many scientists talk about the surface of Titan as being an analog for the conditions of the young Earth billions of years ago, where, despite the frigid temperatures, a rich mixture of carbon compounds could conceivably support the emergence of life. There are even theoretical metabolisms for life in these conditions.[15] For example, instead of respiring oxygen, organisms might take in molecular hydrogen and metabolize it with acetylene (instead of a sugar, such as glucose), or methane could perhaps replace water as a solvent for biological processes.

We don't know whether any of that is actually taking place, but the extremes of Titan's outer crust might also hide a more temperate and familiar inner world. Cassini's measurements of the movement of Titan's land features indicate that the entire outer shell of the moon may be disconnected from its deep interior, floating on an internal ocean of liquid water kept warm by the heat of a rocky core. That ocean, like those of Enceladus or Europa, could provide an incubator for further chemical complexity and life along already known lines. It's just conceivable that Titan is a place of dual ecologies, one on the frigid surface and one in the temperate inside, like some mythological duality of kingdoms.

For all of these reasons, plans are afoot to send a new kind of robotic explorer to Titan, a machine that will fly in Titan's thick atmosphere and low gravity (barely 14 percent of Earth's). This mission, dubbed Dragonfly, is conceived as an "octocopter": a four-pronged rotorcraft, or drone, with double-stacked, meter-sized rotors on each arm that provide lift for a 450-kilogram probe.[16] During the long Titanian nights, which last for some eight Earth days, the drone's batteries will be recharged using electricity generated from the heat of plutonium-238 oxide pellets, enabling flights of many kilometers in distance and altitude. Equipped with cameras, drills, and instruments to analyze the compositions of Titan's environment, Dragonfly could explore land and sea, and conceivably discover the extent to which Titan's hydrocarbon chemistry has evolved in complexity.

If this ambitious mission succeeds, it should arrive at Titan sometime after 2034. At an average distance from the Earth of around 1.4 billion kilometers, it takes electromagnetic radiation more than an hour and twenty minutes to make the one-way journey to Saturn. Consequently, with a nearly three-hour round trip for data and commands, Dragonfly will have to be highly autonomous, making most of its decisions on its own in real time, or else risk encountering catastrophic situations and also being supremely inefficient if it always has to wait for our commands.

Current design specifications call for the octocopter to carry a high-gain antenna for direct communication to the Earth, with no orbital way station. But those long Titan nights also block communications with the inner solar system. During these nights, Dragonfly will have to stay grounded. Although it might gather various data during that time, it's deemed simply too risky to allow the craft to fly around when there is no chance of it checking in for days. Of course, this could change with future advances in AI, and it's a fair bet that if Dragonfly succeeds in its primary three-year scientific mission, engineers will be champing at the bit to test a greater degree of autonomy as time goes by.

Places like Titan, together with the rest of the Saturnian and Jovian systems, aren't just of exceptional scientific interest; they also catalyze fundamentally new ways of exploring space. Their distance, the light-travel time from the Earth, their sheer scale, and the physical spread of attendant moons and natural satellites create new and perplexing challenges compared to what we've contended with around the Earth and the Moon.

The giant planets have forced us to work with our machines in new ways. From Pioneer, Voyager, Galileo, Cassini, and newer generations of probes, we've learned how to operate for decades in space, a very long way from the Sun. We've learned to be willing and able to hand over much more control to our avatar-like probes. Of course, autonomy has been seeping into major facets of our lives for centuries on the Earth. The Industrial Revolution of the West, which began in the late 1700s, started creating more-complex machines to carry out repetitive tasks. In daily life today, we routinely hand off decisions to machine-learning systems, from how we craft text on our phones to what music we listen to. Although self-driving cars and trucks may have been a little slower to perfect than anyone first thought a decade ago, traditional automotive vehicles are already highly augmented with semiautonomous traction controls and antilock brakes, as well as sophisticated autonomous engine and suspension control systems, to create the illusion that we're excellent drivers.

However, space exploration makes the capacity for autonomy absolutely, unequivocally essential while also keeping humans—with our problem-solving abilities and intuitions—firmly in the loop. Autonomy and robotics in space exploration present progressively greater challenges and opportunities than anything on the Earth. The combination of decreasing physical accessibility and increasing causal disconnect because of the finite speed of light is a potent brew for innovation. In a way, the situation for the frontiers of space exploration today is much more like the situation during the voyage of the *Beagle* in the 1830s. Communication then was an arduously slow process, with messages being passed on via land and sea as physical pieces of paper or in one-to-one conversations. Commercial telegraph systems would begin operation only in the late 1830s, and they were seldom end-to-end solutions for sending information. A ship voyage across the oceans required self-reliance and needed meticulously planned supplies and tools for repair and maintenance. And it had to have the right autonomous software embedded in the soft brains of captains and crews.

Jupiter and Saturn are extremely alien places compared to what we've encountered in parts of the inner solar system. Yet in their icy moons, with their wholly unexpected interior oceans, they might actually harbor life that has emerged and evolved along a thread entirely independent from us. In figuring out how to examine and engage with these places, we don't just get to develop systems of autonomous exploration; we also potentially get to discover other inhabitants of the solar system. If we do, the implications will be more than just scientific; they will also be about our moral obligations to other life-forms and our moral obligations to the life on Earth that is, through us, reaching ever farther outward.

Do we treat our prowess at exploration as a supreme privilege, to be exercised with the utmost care on behalf of alien life-forms (as when we dispose of our spent spacecraft in the giant worlds), or is our potential impact on them just part of natural selection? In our recent past, the horrifying application of ideas like Social Darwinism has created pushback and moral certitude that there is no underlying biological justification for many human actions, that they are at best emergent social responses and at worst perversions and unnatural. At the same time, modern scientific views, like the conceptual framework of sociobiology, seek a more nuanced and careful assessment of how the behaviors of organisms can be at least *partially* driven by evolutionary principles and the propagation of genes.

Those tensions, between what we label the natural and the unnatural and what social and moral code we base our actions on, represent a type of boundary right alongside the boundaries of space and time.[17] Together with other kinds of limits and constraints rooted in the physical world, these are key markers for our future story and are what we need to wrestle with next. That's because the idea of Dispersal is one where the sheer scale and scope of life's future extension into the solar system profoundly changes things: not because of some new (and unlikely) cultural enlightenment from within but because of what the enormous expanse of space will do to dilute and change our species and all others.

— 8 —

THE INNERS AND
THE OUTERS

The reef-constructing corals have indeed reared and preserved wonderful memorials of the subterranean oscilla-
tions of level; we see in each barrier-reef a proof that the
land has there subsided, and in each atoll a monument over
an island now lost. We may thus, like unto a geologist who
had lived his ten thousand years and kept a record of the
passing changes, gain some insight into the great system
by which the surface of this globe has been broken up, and
land and water interchanged.

—Charles Darwin,
The Voyage of the Beagle

If you hang around with scientists long enough, especially physicists,
you'll hear talk of "boundary conditions." These might be the measurable extrema of a physical system or the mathematical limits of
an equation that help us fix or constrain the shape of things inside
those limits. The Earth's climate is a good example of a system with

boundary conditions. If—as was done in 1959 with Explorer 7—you measure the total solar power hitting the upper atmosphere and the total power coming back out, you can use these boundary conditions to help pin down the properties of the planet's net climate state and its surface environments. Our solar system and our exploration of space have their share of boundary conditions too, and these conditions are connected to one another. For example, the location and character of the innermost and outermost major planets are critical markers of the history of the solar system's formation and of the pathways taken by planetary evolution, but they also provide a demarcation of our present technological and physical limits for exploration. Those limits are a key part of scoping out the prospects for an interplanetary future and how Dispersal might actually function.

When it comes to precise details of how the cosmos makes planets, our understanding is still shrouded by many mysteries. In the broadest terms, we think that a majority of worlds, particularly the smaller, rocky ones, form the way that the philosopher Immanuel Kant proposed in 1755. Kant's "nebular hypothesis" was that stars and planets come together from slowly rotating, gravitationally contracting clouds of gas and dust from within the interstellar nebulae that are scattered across our galaxy and that are often captured by astronomers in beautiful pictures.[1] Planetary objects condense somehow (Kant waves his hands in the air on this part) from swirling disks of this material that might surround a star like our Sun, as it also condenses out of that same nebular stuff. Modern astronomical data and mathematical physics models seem to confirm this general scheme, albeit with much more colorful scientific detail and nuance than Kant or his contemporaries like Laplace could have suspected. Some of these data, taken using both Earth-bound telescopes and exquisitely sensitive orbital observatories, capture planets and stars in their very embryonic forms. Other data catalog an ever-growing list of mature, fully formed worlds around thousands of stars in our galaxy. These exoplanets show enormous variety in their sizes and orbital arrangements, and as our measurements improve, we can see that these other worlds are also extremely varied in their compositions of gases, liquids, and solids. Each is as unique as any planet in our own solar system.

All of this speaks to how Kant's process is a great cosmic sandbox for planetary formation and history. Those swirling, orbiting materials are subject to a rich interplay between randomness and nature's fixed laws as unpredictable circumstances are corralled and diverted by the boundaries of universally expressed principles of forces and motion. As a consequence, although planetary systems are all readily identifiable as such, they are also almost always different from one another. It's not unlike humans themselves: we're all very obviously the same species, but individually we seldom appear exactly the same.

However, much like the story of Darwin's coral reefs and islands, the complicated tree of tipping points and events that lead to a planetary system makes it very challenging to take what we see today and wind the clock back to understand how things came to be. This is as acute a problem in our own solar system as anywhere else. That's not just a scientific headache; it's a major problem for decoding the function of planetary environments and for prospecting for the kinds of resources and possibilities that a spacefaring species might desire.

That's where the boundary conditions represented by the inners and outers come in—the worlds closer to the Sun than the Earth, and the worlds beyond Saturn. Not only do these planets seem to bear the marks of interconnected upheavals in the solar system's youth more than four billion years ago; they also present some of the most puzzling "what if?" questions about how planets evolve. For example, there is reason to think that the giant worlds of Jupiter, Saturn, Uranus, and Neptune were originally formed much closer to one another and perhaps rather closer in to the Sun. But this orbital arrangement was not entirely stable. A combination of the planets' gravitational interaction with one another and with a vast number of asteroid-like protoplanetary bodies could have led to a wholesale reconfiguration of the giant worlds. In some scenarios, Jupiter may have "migrated" its orbit even farther inward toward the Sun before turning tail and performing a "grand tack" that took it out to its present location.[2] The planets Uranus and Neptune may have done their own migration, but entirely outward, and it's even possible that they swapped their order around the Sun, with Uranus having been the original "outer" giant planet.

These shenanigans would have played gravitational havoc with the smaller, rocky planets closer to the Sun. An original population of these worlds might have been destroyed—flung either sunward or outward, or even broken apart in collisions. Earth, Venus, and Mars might be the reconstituted pieces of those first worlds, and Mercury, well, Mercury might be a poor stranded embryo—a core of what could have become a much larger world in different circumstances. Consequently, the path that these inners and outers were set on billions of years ago has created worlds that define many different kinds of planetary limits. Nowhere is this more obvious than with Venus.

The first up-close measurements of Venus, from the Mariner 2 spacecraft, told us that this world, despite being almost the same size as the Earth and only a third closer to the Sun, was in all other respects as different as Mars is from the Earth. With its 500-degree Celsius (932° F) surface temperature (similar to a very hot brick pizza oven) and crushing carbon dioxide atmosphere with over ninety times Earth's surface pressure, Venus is a hellish place. In these conditions, aluminum and glass can melt, and metals like pure titanium can even combust, making Venus a whole other proposition for exploration.

From the 1960s to the 1980s, the Soviet Union was the most prolific and ultimately successful nation when it came to exploring Venus. This seems to have resulted from a mixture of luck and expediency. Most of the first ten Soviet attempts to send spacecraft to Venus didn't succeed at all, and there were those notable early missteps in determining Venus's actual location. But with the United States' very public successes at the Moon and Mars, Venus was a distinct opportunity for the USSR to demonstrate its space prowess. It didn't hurt that Venus was also a popular subject in Soviet science fiction of the time, including in movies like 1962's *Planet Bura (Planet of Storms)*.[3] In this fiction, just as with Mars, Venus was speculated to be vastly more hospitable than it actually is, complete with a population of stop-motion dinosaurs.

The first fully successful Soviet mission to Venus was Venera 4 in 1967. This was also the very first time that any probe had directly sampled the atmosphere of an alien world and relayed its discoveries back to the Earth. After a launch from the Baikonur Cosmodrome in Kazakhstan, the mission headed for a four-month cruise down the Sun's gravity well to intercept Venus. Once there, the main carrier spacecraft dropped off a sphere-like, meter-sized capsule that plunged into the Venusian upper atmosphere. After a blistering frictional slowdown in the atmosphere, the battery-powered probe opened its parachutes at an altitude of about 52 kilometers and descended for 90 minutes. During this descent it sent back temperature and pressure data, measured its altitude with a radar instrument, tracked the humidity (there was only a trace of water), and performed an analysis of the atmospheric composition. All of this information was pulsed back via radio signals relayed to the Earth by the carrier craft, at a painfully slow 1 bit of data per second.[4]

The Venera 4 capsule didn't make it to the surface intact, though. Later analyses indicated that somewhere in the depths of the atmosphere, the probe, which was pressurized to only twenty-five atmospheres, most likely succumbed to the external pressure and was crushed—or simply ran out of power. But its fate didn't really matter because of the fantastic data that were returned. Venera 4 confirmed that Venus was bone-dry and scorching hot in its lower atmosphere and that this atmosphere was 96 percent carbon dioxide, a few percent of nitrogen gas, and a trace of other compounds. The probe also discovered that Venus's opaque clouds are mostly droplets of sulfuric acid. We now understand that this acid is formed by sunlight triggering chemical reactions in the atmosphere's complement of carbon dioxide, sulfur dioxide, and what little water vapor there is.

Buoyed by this success, more Venera missions were developed and flown. Launched just five days apart in 1969, the Venera 5 and 6 probes—each very similar in design to Venera 4—also successfully entered the Venusian atmosphere and sent back even more data. Then, in 1970, the Venera 7 probe—a beefed-up version of the earlier Veneras—actually made it intact and "live" onto the surface of Venus despite its parachutes failing partway down, causing the probe to smack into the ground at about 59 kilometers per hour.

At first, the mission scientists thought that Venera 7 had simply been destroyed, for its signals seemed to abruptly end when it hit. But radio observers on Earth had fortunately kept their data recorders running and, a few weeks later, discovered 20 minutes of weak signals sent by the probe after it had landed. It seemed that the capsule had bounced on impact and ended up on its side, with its transmitting antennae misaligned with the Earth in the sky above it. Those 20 minutes of data finally confirmed just how intense the pressure and temperature are right at the planet's surface and made Venera 7 the very first human probe to "soft land" (in the euphemistic language of space exploration) on another planet.

In 1972 the Venera 8 entry probe got everything exactly right. It was prechilled by its carrier spacecraft, and it descended smoothly to the hot surface, where it functioned for 50 minutes. Not only did this probe's photosensors discover that the Venusian cloud layers give way to a clear lower atmosphere; they also determined that there was enough light on the surface to send cameras to take images. That paved the way for Venera 9—a hefty 5-metric-ton spacecraft launched in 1975.

Figure 8.1. The Venera 9 lander. The lower ring structure was designed to partially crumple on impact. (Credit: NASA/NSSDCA.)

This Venera consisted of both an orbiter and a more advanced lander packed inside a 2.5-meter-diameter sphere for transit. The lander itself was an iconic, brutal-looking thing designed to withstand both the Venusian environment and a crunching impact on the surface at 25 kilometers per hour. When Venera 9 arrived, it became the first mission to orbit Venus and the first-ever mission to send back an image from the surface of another planet. (See Figures 8.1 and 8.2.)

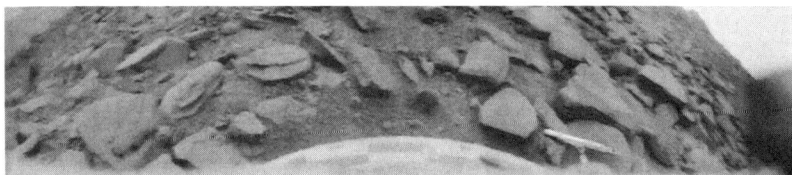

Figure 8.2. The first-ever image from the surface of another planet. Venera 9's partial fish-eye television-camera view of Venus showing rocks and finer soil, the edge of the lander itself, and (on the upper left) a more distant hill or rise in the Venusian landscape. Surface lighting was similar to a cloudy summer day at middle latitudes on the Earth.

From here on, the Soviets undertook one of the most ambitious and successful campaigns in planetary-exploration history, a campaign that is, to this day, often overlooked by the rest of the world. Following Venera 9 there were *seven* more Venera missions, almost all entirely successful. These included five more landers—sending a wealth of data back to Earth—and three orbiters, as well as four fly-bys of Venus by the carrier spacecraft. During this period, the only other exploration of Venus was carried out by the United States' Pioneer Venus 1 and 2 missions in 1978. These were also ambitious and successful, with their own scientific focus. On Pioneer Venus 2, a suite of four atmospheric probes of differing sizes were sent into the Venusian clouds, scattered across the planet's nightside and dayside, with one surviving just long enough on the surface to send back data.

In 1984 came what would be, in retrospect, the last and quite spectacular hurrah for Soviet Venus exploration. This mission consisted of two large mother ships, Vega 1 and Vega 2, launched within days of each other and arriving at Venus in June 1985, just three months after Mikhail Gorbachev became the final leader of the Soviet Union. This pair of spacecraft was developed out of a collaboration between the Soviet Union and eight other countries in Europe and the Soviet bloc. Each flew by Venus, and each dropped off a lander and an innovative atmospheric probe—the likes of which had never been implemented before. After this drop-off, the twin mother ships then used Venus for a gravity assist to intercept Halley's Comet in 1986 as this famous object made its once-every-seventy-six-years periapsis swoop through the inner solar system. In that encounter, Vega 1 returned the very first up-close images of the comet's dirty nucleus of rock, dust, and water ice.

Before all of that, back at Venus in 1985, the two Vega landers parachuted their way to the surface, touching down in the Venusian nighttime to gather data. For these landers, Soviet engineers had devised an ingenious system to let them drill into the soil and move a sample inside the craft in order to study its composition using X-ray fluorescence. Given the extreme pressures and temperatures on the planet's surface, this was a risky experiment—both in terms of making a functional drill and by allowing a sample into any part of the lander. To handle this, the engineers fabricated a drill that would function only after the heat of the Venusian surface had caused its parts to expand, and they designed an innovative miniature airlock for drawing the hot soil sample into the pressurized probe.

Meanwhile, the third and most audacious component of the Vega missions was a pair of balloon-borne capsules, or "aerostats" (sometimes called "aerobots").[5] These probes were carried downward by the landers until they had slowed sufficiently in the atmosphere. At an altitude of around 54 kilometers, each detached from the plummeting craft. They then released and inflated helium balloons covered in Ftorlon fabric (a type of Teflon-like material) that were about 3.5 meters across. These balloons, designed to withstand the corrosive sulfuric acid of the clouds, each supported a small gondola of scientific instruments that dangled underneath on a 13-meter-long tether.

This was not a gentle ballooning excursion, though. Although the pressures and temperatures at this altitude are surprisingly similar to those on the surface of the Earth, the winds are hurricane force, and the aerostats flew along at 250 kilometers per hour, each probe sweeping from the nightside to the dayside of Venus over the course of a journey that encompassed some 30 percent of the planet's circumference. After 46 hours of this wild ride, the probes' batteries ran out.

The hair-raising circulation speeds of Venus's atmosphere had been known since the 1960s, when Earth-based radar measurements peered down to the surface to gauge the rotation of the solid body. To everyone's surprise, the rocky surface turned out to rotate only once every 243 Earth days, and it does this in the opposite sense to Venus's orbit around the Sun in what's known as a retrograde rotation. But the atmosphere of Venus is tearing around sixty times faster, completing a circuit in only 4 Earth days. This is called super-rotation, and there is evidence that the thick atmosphere should exert enough force on Venus's elevated hills and mountains to actually speed up the rotation of the solid planetary mass by around two minutes for every Venusian day if it is applied consistently. That presents a conundrum, though, because it wouldn't take long for that acceleration to force the planet to speed up, so there must be opposing forces that act to slow or dampen those changes. It seems that Venus may be endlessly speeding up and slowing down by small amounts, like a sailing vessel caught in an inconstant trade wind.

For planetary scientists, Venus remains a very puzzling place. The biggest unanswered question is simply why did this world turn out the way that it did? On the face of it, Venus and the Earth are remarkably similar in size and proximity to the Sun, yet one is an arid pressure cooker with acidic clouds, and the other is a temperate world of water oceans and land.[6] We still don't know the answer, but it's possible that rocky planets like these always encounter many turning points in their histories, times when the roll of the cosmic dice can send a world onto one or another path. For Venus, it is possible that the timing of past volcanic activity and the mechanics of its outer rocky crust coincided with a set of atmospheric properties and climate responses that pushed it into a one-way street toward its present super-greenhouse condition. Some of my scientific colleagues have generated virtual Venusian worlds, using sophisticated climate simulations and models of geophysics, to suggest that the young Venus could have easily been replete with shallow water oceans and land. But that water might have sealed Venus's fate if ill-timed volcanism also thickened the atmosphere and raised temperatures. This would have allowed a steamy greenhouse to—in effect—dry the planet out by pushing water into the stratosphere, where its elements would be lost to space.

Venus remains an enigma, even though orbital missions that came after the Vega probes have yielded global radar maps of Venus's surface and more details of the atmosphere and its cloud physics.[7] Today, there are hopes for new spacecraft and new opportunities to decode what has made Venus the way it is. These include the first privately funded small probe to Venus, designed to scrutinize the planet's atmospheric contents and look for unexpected molecules like phosphine that could, just conceivably, reflect the presence of alien, cloud-living microbes.[8]

The present state of Venus has served our species well, though, even if that seems like an odd thing to say. As with the Moon and Mars, Venus has its own unique exploration challenges that force us to innovate and to stretch our technology and imaginations. For example, the aerostats of the Vega mission might have been a last gasp of Soviet-era ingenuity, but they were also a proof of concept of future options for a presence at Venus. Conceptual design studies have been made for large floating laboratories in the more temperate zones at high altitudes in the planet's atmosphere, and, looking to the more distant future, there has been serious discussion of the possibility of placing human-rated buoyant habitats in the Venusian atmosphere. These floating platforms would be situated within the very manageable external pressures and temperatures that exist just above 50-kilometer altitudes, with "aeronauts" experiencing gravitational acceleration that is a healthy 90 percent of that on Earth. Although Venus is an example of the boundary conditions for hospitable terrain, there are still parts of it that humans and machines might be able to experience.

However, to move closer to the Sun than Venus is to begin to touch the barriers to current exploration possibilities. If Venus is a planet that might have been something else but for the roll of geophysical dice, then Mercury is a world that never really made it to the starting line. The most striking global property of Mercury is that it appears to have an iron and nickel core that is much larger in proportion to its mass than that of any of the other rocky worlds. This core takes up a full three-quarters of the planetary mass and is wrapped in a silicate rock mantle and heavily cratered outer crust. We don't know what led to this. It's possible that Mercury, similarly to the proto-Earth, experienced a huge protoplanetary collision that stripped away the original silicate layers, leaving this metal-rich composition. Or, this close to the Sun, the way that Mercury agglomerated out of gas and dust was fundamentally different than for more distant planets, causing a preferential accumulation of heavy elements.

Mercury also has a vexing rotation and orbit. It takes 88 Earth days to orbit the Sun in a path that is the most elliptical of all the major planets—coming 34 percent closer to the Sun at perihelion than at aphelion. Mercury also rotates only once every 58 Earth days, which means that the actual length of illumination on the surface— a solar day—is 176 Earth days, from sunrise to sunset, followed by the same length of darkness. Because of this, the pace at which night or day crosses the equatorial surface is only 3.5 kilometers (about 2.2 miles) per hour, roughly human walking speed and far slower than the 1,600 kilometers per hour at which twilight sweeps across the Earth's equator. You could, if you had the supplies and stamina, experience permanent daylight (or night) on Mercury by simply walking westward all the time, although that might be challenging for other reasons: the daytime temperature on the surface at the equator can range from 427° Celsius when the planet is closest to the Sun to around 280° Celsius at aphelion. Night can be a chilly −163° Celsius, and at the poles where—as on the Moon—craters offer zones of permanent shadow, the temperature can stay below −170° Celsius at all times. In 2012 it was confirmed that the shadowed area of Mercury's northern pole has large water-ice deposits right at the surface. This is probably the closest permanent water ice to the Sun anywhere in the solar system.

Getting to Mercury from the Earth is, perhaps counterintuitively, one of the most difficult maneuvers for interplanetary missions.[9] Mercury is so deep in the Sun's gravity well that its orbital speed is close

216

to 48 kilometers per second—nearly 18 kilometers per second more than the Earth's. Any mission to orbit the planet or land softly has to find that Delta-v difference somehow. The first successful probe to Mercury—Mariner 10, launched in 1973—actually avoided some of this cost by using a gravity assist from Venus to drop the perihelion of its sun-centric orbit enough to let it pass close to Mercury in three brief encounters that ended in 1975, leaving the spacecraft to orbit the Sun forever. To function this close to the Sun, Mariner 10 had to withstand five times as much solar-particle radiation as experienced near the Earth. It needed extra shielding to maintain its electronics at cooler temperatures, and because its solar panels needed to stay below 115° Celsius in order to work properly, it had to keep these arrays tilted at an angle to the Sun to limit how much light they'd absorb.

It wouldn't be until thirty years later, in 2004, that a mission launched for Mercury with the goal of actually orbiting this world. That spacecraft was NASA's MESSENGER (a name born out of obsessive acronym creation: MErcury Surface, Space ENvironment, GEochemistry, and Ranging). The mission also called for a convoluted journey that consisted of an Earth flyby and gravity assist a year after launch, two Venus gravity assists in the next two years, and then three Mercury flybys. Then, in March 2011, the spacecraft was finally able to sidle up to Mercury at a slow enough relative speed to use its large rocket engine to transfer to a highly elliptical twelve-hour orbit around the planet. This orbit was designed to keep the spacecraft from spending too long over the dayside of Mercury. Although MESSENGER had a strongly reflective shade of curved ceramic fiber to protect it from the Sun on one side, it still needed to expose its instruments to the planet. But the heat and reflected light coming from Mercury's surface itself were enough to damage the probe. By swooping in at high periapsis speeds and then cooling off at apoapsis, MESSENGER could avoid getting too hot, not unlike the orbital strategies used at Jupiter to avoid radiation damage.

MESSENGER also did something extremely clever to reduce the fuel it needed for small maneuvers: it used the pressure exerted by the Sun's light to adjust its trajectories. This may seem strange because it is a property of electromagnetic radiation that we don't experience in normal day-to-day life on the Earth. That's because it takes an awful lot of photons to cause enough of an effect for our senses to register. Nonetheless, we've known about the phenomenon since 1862, when the physicist James Clerk Maxwell codified the fact that light carries momentum and, consequently, that light must exert a mechanical pressure when it interacts with a surface.[10]

Therefore, when objects absorb or reflect light, they can be accelerated by an exchange of momentum, just as with the exchange of momentum in rocketry. The size of acceleration is very modest in most terrestrial situations. At the location of the Earth, raw sunlight exerts the same pressure as a milligram of mass spread across a square meter in Earth's surface gravity. This is tiny in human terms, but it can make a significant difference in space, where small forces add up over time. In fact, the pressure of light is an important consideration for understanding the long-term orbital and spin behavior of objects like asteroids. Not only does sunlight push on objects when it is absorbed or reflected from them; when it is absorbed, it also heats up the sun-facing side of objects' surfaces. That heat is then lost as the reradiation of thermal, infrared photons, causing an additional thrust from the object itself. Together, these components of force produce what are called the Yarkovsky effect and the YORP effect, the former changing asteroid orbits over millions of years and the latter changing their spin and orientation.

At Mercury's distance from the Sun, the solar-radiation pressure is about eleven times larger than at the Earth. When Mariner 10 maneuvered around the Sun in the 1970s, engineers were able to adjust its orientation by tilting its 5 square meters of solar panels to get the right "push." MESSENGER upped the ante by tilting its panels and using its sunshield to adjust its course trajectory throughout its entire mission—using, in effect, a "propulsion-free trajectory control" much in the way that sailing ships use the winds and the oceans. This meant that MESSENGER's six planetary gravity assists could be tuned precisely with minimal fuel use. In the end, and rather poetically, when MESSENGER did run out of fuel to maintain its orbit around Mercury, it was the force of sunlight that drove the orbital changes that eventually took the spacecraft to an abrupt and destructive end on the planetary surface in 2015.

If Venus and Mercury represent one set of limits on exploration and on our understanding of planetary history, then the outer worlds are there to remind us that we're still rank amateurs. Take, for example, Uranus and Neptune: these two ice-giant worlds represent a class of planet that—based on what we currently know about exoplanets—appears to be extremely common across our galaxy and perhaps the universe as a whole. With masses between fifteen and seventeen times the mass of the Earth, they are a fraction of the size of the gas giants but enormous compared to the rocky inner worlds, each measuring about four times the diameter of the Earth.[11]

At distances of 3 billion and 4.5 billion kilometers from the Sun, respectively, Uranus and Neptune experience illumination more than 330 and 880 times fainter than we experience at the Earth. Their orbital years also amount to 84 and 165 Earth years. In fact, since Neptune was first fully documented with a telescope by astronomers in 1846, it has completed only a little more than one full orbit of the Sun. Its northern hemisphere has been in summer since 2005 and will continue to be so until fall begins around our year 2045.

We've visited Uranus and Neptune precisely once, with the Voyager 2 probe. This mission first traversed the Uranian system, and its complement of twenty-eight confirmed moons, at a distance of 50,000 kilometers, and then traversed the Neptunian system three years later, in 1989, when the spacecraft passed within just 5,000 kilometers of the planet, well within the orbits of Neptune's sixteen known moons. Since that time, our only new information about these planets has come from telescopic data from the Earth, as well as from space telescopes like the Hubble and the JWST.

Where these ice giants most differ from the gas giants is in the size of their outer layers. For Jupiter or Saturn, more than 80 percent of the planetary mass is in the form of hydrogen and helium, but for Uranus and Neptune only 10 percent to 20 percent of their total mass is in this form, with the rest presumed to be in solid or molten heavier elements in their interiors.

This far from the Sun, the surface temperatures of these worlds are so low that most of the compounds that can condense to form clouds actually do so well below the visible gaseous exterior, leaving these planets with rather bland appearances—except when material wells up from beneath. Because we have so few in situ data on Uranus and Neptune, their deep interiors are little understood, except through models (and boundary conditions). They probably have hot rocky cores of silicates and iron-nickel alloys that are surrounded by water ices at very high pressures and temperatures. But there may not be distinct layers throughout these worlds; instead, the various components may blend into one another. Given these factors, it has been suggested that the term "ice giants" is really not terribly appropriate; these planets are more like tepid "subgiants" or "outer giants." They're also quite different worlds from each other. Uranus seems to radiate only a little more energy than it absorbs from the Sun, but Neptune radiates over twice as much as it absorbs (indicating that it has high interior temperatures). It's often assumed that the internal energy of these planets, as with the gas giants, comes from their inevitable and gradual gravitational contraction—the conversion of potential energy into kinetic energy in their gas and fluids—which would make Uranus an oddball. Perhaps Uranus is just an older world than Neptune, or perhaps something disrupted its interior and sped up the release of heat long ago. Or maybe this world simply unleashes its pent-up thermal energy sporadically—we don't yet know.

Uranus has another special external feature: its spin axis is enormously tilted, putting it almost on its side relative to the plane of its orbit. Depending on how you choose which end of the planet represents north, it can also be thought of as having a retrograde spin—in the opposite sense to its orbital motion. Like some giant sideways-spinning top, this means that for half of its orbit around the Sun, one pole is illuminated and then the other. Each pole gets forty-two years of continual sunlight followed by forty-two years of continual darkness. Planetary scientists have been trying to explain this sideways configuration since 1787, when the astronomer William Herschel measured the orbital orientation of two of the planet's moons and deduced Uranus's axial orientation. A favored idea is that three or four billion years ago, a planet larger than the Earth collided with Uranus and pushed its axis over, disrupting its interior.

Neptune also has its share of features and puzzles. It has many more visible storms in its upper atmosphere, including a selection of dark-featured "spots" that seem to come and go over the span of a few Earth years. Although it has sixteen moons, almost 99.5 percent of the mass of those satellites is contained in just one object—the moon Triton. But Triton appears to be an interloper that formed elsewhere in the solar system.[12] Unlike a moon that had formed in situ, Triton orbits Neptune in retrograde—against the sense of the planetary spin—and this causes gravitational tidal forces that act to shrink its orbit (unlike our own moon, where forces are expanding its orbit). Triton is actually on a slow, multibillion-year death spiral down to the planet. Eventually, it will reach the Roche limit, which we encountered earlier, and be disrupted and torn into a temporary ring of material around Neptune by tidal forces. The moon's composition, with a translucent outer sheath of frozen nitrogen on top of a crust of solid nitrogen, water, and carbon dioxide, also points to an origin farther away. It is almost certainly a captured world from the outer, outer solar system, from the zone we call the Kuiper belt.[13]

The Kuiper belt is a little like how the inner asteroid belt between Mars and Jupiter would be if it took a course of bodybuilding steroids. It's a zone of primordial objects orbiting the Sun beyond Neptune where more than 100,000 of these are larger than 100 kilometers across, and there may be trillions of smaller, icy bodies. These objects' orbits spread out nearly twice as far from the Sun as Neptune's orbit, and this cosmic population is full of surprises. The first of those Kuiper belt surprises, although nobody knew this at the time of its discovery in 1930, is Pluto.

In 2015 a probe called New Horizons flew through the Plutonian system at 12,000 kilometers above Pluto's surface and captured the first up-close images and data on this system.[14] Because the primary mission of New Horizons was to get to Pluto, with only one other encounter at Jupiter for a gravity assist, there was a gap of nearly ten years between its launch and its major science operations (except in 2007, with the Jovian encounter), and the probe was in electronic hibernation for most of that interplanetary journey. For the human scientists and engineers, this meant nearly a decade of doing other things—carrying on with the ups and downs of their professional and personal lives while New Horizons cruised away from the Sun.

When it finally woke up at Pluto, New Horizons sent spectacular confirmation that this is a very different type of world: a small, ice-rich object that is so far from the Sun that its surface is caked not with water but with a patchwork mix of frozen nitrogen, methane, carbon monoxide, and other compounds. Beneath a thick crust, its interior may harbor yet another liquid-water ocean. The Sun's gravitational tides are tiny this far out in the solar system, and complex orbital hierarchies are not unusual because they're less likely to be rendered unstable. As I've mentioned before, Pluto is separated by only 20,000 kilometers from its binary world companion Charon, and these bodies waltz around a point in empty space between them every six days, whereas the small satellites Styx, Nix, Kerberos, and Hydra all orbit around that Pluto-Charon binary in periods that range from twenty to thirty-eight days, making this a veritable swarm of objects that could not exist in the inner solar system.

Objects in the Kuiper belt are, however, strongly affected by Neptune's gravitational pull, and in the range where orbital periods are in whole-number ratios with Neptune's orbital period (just like the so-called mean-motion resonances of the Galilean moons around Jupiter), there are significant consequences. For example, Pluto comes close to a 2:3 ratio, with two Pluto orbits for every three Neptune orbits. Pluto's orbit is also inclined and significantly elliptical, and because of that it actually crosses the orbit of Neptune, moving inside that radius for 20 years every 248 years (last doing this from 1979 to 1999). But because Pluto is locked in the 2:3 resonance, it actually never comes close to colliding with Neptune or being gravitationally scattered because the clockwork-like resonance ensures that the two don't meet. An explanation may be that in the young solar system, the orbital evolution of Neptune involved exchanges of momentum with trillions of smaller bodies orbiting the Sun. These exchanges would have "pushed" some objects into mean-motion resonances, and Pluto may have just gotten lucky.[15]

Pluto used to be considered the outermost and last major planet. But in the last several decades, astronomers have detected many similarly large Kuiper belt objects and many farther-placed bodies with the broad label of trans-Neptunian objects (meaning beyond Neptune), all arrayed across many distances from the Sun.[16] These include 1,000-kilometer-wide bodies like Quaoar, with its own moon and a double ring system.

Naming conventions of these objects reflect the history of human science and evolve with social change. In the twenty-first century, most trans-Neptunian objects are named after multicultural creation deities, with Quaoar associated with the Tongva people, who are indigenous to the Los Angeles basin. Another large dwarf planet is Eris (goddess of discord in Greek mythology), a world that is roughly the same size as Pluto at 2,000 kilometers across, has a moon, and has an extremely elliptical 560-year orbit that takes it out to 97 times farther from the Sun than the Earth is. But that orbit is not atypical, and another trans-Neptunian object beyond the Kuiper belt, called Sedna (a sea goddess in Inuit mythology), has an outer orbital aphelion that is 937 times farther from the Sun than is the Earth, with a civilization-spanning orbital period of more than 11,000 years.

After the New Horizons probe zipped unscathed through the Plutonian system, an audacious plan was put into play to try to find another Kuiper belt object to intercept. But because the probe had only limited propellant left after encountering Pluto, it couldn't deviate from its trajectory by more than a fraction of a degree. And because of communications becoming more and more difficult, it needed to find a target no farther than 55 times Earth's distance from the Sun (bearing in mind that Pluto is already 34 times farther from the Sun). After an intensive search, in 2014 new data from the Hubble Space Telescope revealed three potential destinations. On January 1, 2019, New Horizons—still moving at more than 14 kilometers per second relative to the Sun—flew just 3,500 kilometers from a surreal-looking, reddish-brown dumbbell-like object, 36 kilometers in length, called Arrokoth (representing the Powhatan peoples and shown in Figure 8.3).

This encounter represents what is, for now, the limit of active planetary exploration in the outer solar system. It took more than twenty months for New Horizons to return its data from Arrokoth in a technical chore that highlights another kind of critical boundary condition. Apart from the forbidding amount of time it takes to travel to the outer planets and beyond, and the reliance on independent energy sources like radioisotope-powered thermoelectricity, these are also extremely challenging distances and places from which to send and

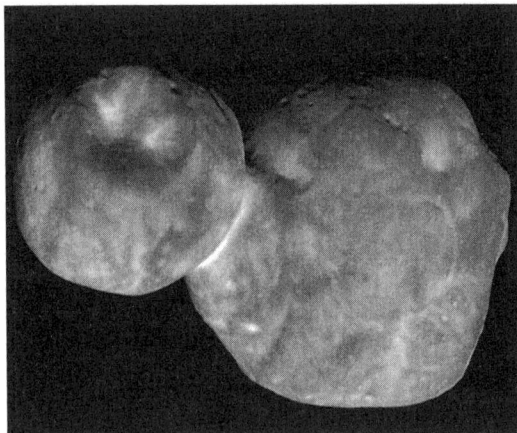

Figure 8.3. Arrokoth is an ancient Kuiper belt object intercepted by the New Horizons mission and is the farthest object from the Sun to which we have ever flown this close: a distance of 3,500 kilometers. (Credit: NASA/Johns Hopkins University Applied Physics Laboratory/Southwest Research Institute.)

receive signals. The New Horizons probe captured just over 6 gigabytes of data during its 2015 encounter with Pluto. That's about the same quantity of data that would be in a two-hour high-definition streaming movie. But the New Horizons data took more than sixteen months to trickle back, at a rate of only 2 kilobits per second, reminiscent of the era of dial-up modems. To achieve even that painfully slow datastream took a 2-meter-diameter high-gain antenna on the spacecraft and the full attention of the Earth's Deep Space Network, which includes three geographically separated 70-meter radio dishes: one near Goldstone in the Mojave Desert, one near Canberra in Australia, and another near Madrid in Spain, combined with additional clusters of 30-meter-size dishes at each global site.[17]

The problem is that a deep-space probe like New Horizons can spare only 12 watts of power for its transmitters, and its beamed radio emissions are spread out and diluted across the intervening billions of kilometers. Speaking to the probe or capturing its returned data involves big, sensitive radio telescopes and a lot of electronic processing to eke out the signal among the noise. It's an immense task for anyone, and the European Space Agency also has its own dedicated deep-space communications network, as do the space agencies of China, Russia, Japan, and India.

For the even more distant Voyager 2 probe, the power of its transmissions now collected at the Earth amounts to less than a billionth of a billionth of a watt. For comparison, the average power use of a single human cell has been estimated as a trillionth of a watt, which is a million times more power (although not all concentrated in a pulse of radio photons at a specific frequency). Conversely, in 2023 a set of broadcast commands inadvertently caused Voyager 2's antennae to point two degrees away from the direction of the Earth, preventing normal communications. It took the transmission of a 100,000-watt radio signal from the Earth to get through to the probe and correct its posture.

Planetary positions around the Sun can also cause issues for communications. Whenever the Earth is on the side of the Sun opposite to a spacecraft, communications can be temporarily disrupted. For an inner world like Mars, this causes thirteen-day-long communication blackouts with the Earth roughly every twenty-six months, which is an issue for robots and astronauts alike. For Mercury and Venus, similar lengthy blackouts or periods of poor communications also exist, and they include times when these worlds are simply too close to the noisy Sun in our terrestrial skies. For the outer solar system as well, the farther a spacecraft is from the Sun, the smaller the angular separation of the Earth and Sun, and the harder it becomes to point an antenna at the Earth and not get solar interference.

Testing our technology and skills at the inners and outers helps define what we might call the boundary conditions of a "zone of easiest exploration" for our species. Those conditions include the cost of climbing up and down the Sun's gravity well and the availability—or not—of gravitational assists from the planets. It's true that closer to the Sun, the shorter orbital periods of worlds like Mercury or Venus make useful alignments for spacecraft journeys much more frequent. But then again, we're deeper in Delta-v debt this near to our star. Jupiter may be an excellent source of gravity assists for the outer worlds, but it takes ten years for it to trundle around its orbit, so useful alignments for spacecraft require a lot of human patience.

Near the Sun, solar photons are an abundant source of power, and radiation pressure offers zero-propulsion spacecraft maneuvering. But these same photons can easily overheat probes, and they can even make the surface of a world like Mercury a hazard from orbit. Venus may not be such a threat when orbited, but it's a beast of heat, pressure, and corrosive acidity inside its atmosphere. Out at Uranus and Neptune—and beyond—there is not enough sunlight to power much of anything, and communications are not just delayed by hours; they can also be intrinsically weak and suffer from exceedingly slow throughput rates. To top off all of these factors, human lifespans and terrestrial economic dynamics introduce other kinds of limits. These make it hard to synchronize lives and aspirations with the years and decades needed for spacecraft development and for long interplanetary journeys.

The specifics of the voyage of the *Beagle* were constrained or bounded by a distribution of continents and oceans that had been determined by plate tectonics and billions of years of Earth's geophysical and biological evolution. In exactly the same way, the voyages of space are dictated by the contingencies of gravitational forces and planet formation that—often violently and randomly—sculpted the solar system into its present form. Different planets in different places would have meant a different set of boundary conditions and different zones of easiest exploration. Some exoplanetary systems around other smaller and fainter stars are extremely compact compared to ours, with eight major planets in orbits that would fit inside the orbit of Mercury. A species that evolved there might find space exploration far less challenging. Equally, they might gain far less instruction from their wanderings. Would Darwin have accumulated the same evidence or been stimulated in his thinking the same way had the Earth been different?

Of course, we can't know the answer to that. But the environment and boundaries of exploration dictate the challenges that any species faces, as well as the coping strategies and ideas that it eventually generates, which in turn fuel evolution's experiments and sometimes lead to the invention of entirely new forms of existence. Although we've gauged some of the planetary limits to our zones of exploration, it turns out that our parent star also has some contributions to make and some tricks up its flowing sleeves to keep us guessing, as we'll now see.

— 9 —

THE FURNACE

White massive clouds were piled up against a dark blue sky, and across them black ragged sheets of vapour were rapidly driven. The successive mountain ranges appeared like dim shadows, and the setting sun cast on the woodland a yellow gleam, much like that produced by the flame of spirits of wine. The water was white with the flying spray, and the wind lulled and roared again through the rigging: it was an ominous, sublime scene.

—Charles Darwin,
The Voyage of the Beagle

Something extraordinary began happening to the planet Earth on August 28, 1859, only three months before Darwin's *On the Origins of Species* hit the bookshelves as an instant bestseller. This planetary event had nothing to do with Darwin, though. All across the globe the skies started to blaze with the light of auroras—the glow produced as electrons and protons from the solar wind flow into the Earth's upper atmosphere. These fast-moving, energetic solar particles collide with stratified atmospheric compounds like oxygen and nitrogen, and they set in motion a process similar to what happens inside a confined neon light. Those collisions transfer energy to make temporary ions and excited atomic states that, in turn, produce a cascade of recaptured electrons and resettling atomic-energy levels. They regurgitate energy as photons, producing the red, green, blue, and purple hues of auroral light. To a lesser or greater degree, these kinds of processes happen all the time in our atmosphere. But the auroras in 1859 had an extent and intensity the likes of which few, if anyone, alive at the time had ever witnessed.

A few days later, on September 1, a British astronomer called Richard Carrington was routinely studying the Sun when he saw what he described as a "singular appearance."[1] Carrington worked at an observatory in southern England, where a telescope projected the Sun's disk onto a screen so that he could trace out its patterns of light and dark. As he sketched away, he suddenly saw two searingly bright ovals of white light appear at the location of an enormous 100,000-kilometer-wide sunspot on the Sun's outer surface.[2] By the time he had scurried off to find another witness, these flashes had almost vanished. But seventeen hours later, the consequences for the Earth became very apparent.

The auroras of the preceding days had been exceptional, but now, from the highest latitudes all the way down to equatorial zones in Colombia and parts of Australia, the skies were lit up even more brilliantly. Astonishing curtains of changing, undulating green, red, and blue light filled the heavens as the solar wind engaged with the Earth's magnetic fields at the farthest reaches of the stratosphere. By the second day of these auroras, at 1 a.m. in the Rocky Mountains of North America, the skies were so bright that gold miners thought it was dawn and mistakenly got up and made their breakfasts. Electrical telegraph systems, in use around the world since the late 1830s, became noisy and erratic. Compasses became unreliable, and sensitive devices used to carefully measure Earth's magnetic field shot off the scale. Eventually, some telegraph systems began giving their operators electric shocks, and other telegraph stations were reported to have actually burned down after electrical sparks started fires. Where brave operators were able to gingerly handle their equipment without being stunned, they discovered that it still worked even when they completely disconnected the batteries that usually provided the electrical currents for sending signals.

After a couple of days, all the brilliant auroras and strange electrical phenomena had quietened down. But the ramifications of these spooky events and auroras were longer lived. This episode provided critical evidence that helped refine scientists' emerging grasp of planetary and stellar magnetism, and of the Sun's fundamental nature. We now understand that the "Carrington Event" of September 1 was the result of a massive solar storm. The intense flashes that Carrington saw on the Sun were associated with two giant eruptive magnetic bubbles that entrained material as they exploded away from the Sun's visible surface—its photosphere—and barreled across space to crash across the Earth like a tsunami.

In the grand scheme of things, being on a planet this close to a star is not the smartest move. We orbit an unenclosed thermonuclear system weighing a thousand trillion trillion tons that has been fusing elements continuously in its core for the past 4.5 billion years. This system converts some of its mass into fearsomely intense gamma-ray radiation that scatters and diffuses outward for a hundred thousand years before emerging as the sunlight that we see. At the Sun's outer photosphere there are endlessly churning, hot, electrically charged particles like protons and electrons entwined with vast magnetic fields that flex and snap like angry serpents. Sometimes, this dynamic environment generates streamer-like solar flares of accelerated material, and sometimes it generates giant magnetic bubbles that propel a billion or more tons of starstuff at high speed out into the solar system. If a planet happens to get in the way, it will know about it. These great bubble storms, known as coronal mass ejections, were part of what Carrington excitedly saw in 1859 and were what swamped the Earth shortly after.

When these roiling mass ejections crashed into the Earth's own magnetic field and upper atmosphere in 1859, they encountered a reactive and pliant environment. The angle of attack of the Carrington Event meant that when the first bubble arrived, the solar material flowed fairly harmlessly around and past the planet. But as time went by, the angle changed, and that was when all hell let loose. By the time the second great ejection hit, it compressed Earth's entire magnetosphere—the volume of space where the planet's magnetic field usually resists the pressure of solar wind—squeezing it down from a size of 60,000 kilometers all the way to the stratosphere at an altitude of 50 kilometers. The Van Allen radiation belts (which I'll describe shortly) were wiped away, and raw solar particles smashed down into Earth's atmosphere—setting off more auroras and even triggering a cascade of chemical reactions. We have now found evidence of excess nitrates trapped in air bubbles in the ice of Antarctica that are a direct result of the impact of this particle radiation on Earth's nitrogen-rich atmosphere. The 1859 nitrate excess may be the largest in the past five hundred years.[3]

As the various components of the 1859 storm hit, the Earth's very outermost atmosphere—the exosphere above 500 kilometers—was also heated by X-rays, inflating it by hundreds of kilometers in altitude. The particle processes that drove the generation of nitrates would have also produced a flood of neutrons at Earth's surface. The ozone layer in the stratosphere is thought to have been partially depleted, and it may have taken years to recover. The flood of charged particles flowing along magnetic fields also acted like a giant induction motor, driving powerful electrical currents in the rocky underpinnings of Earth's continents. The geological nature of North America makes it particularly susceptible to these geomagnetic currents, and some ended up leaking into telegraph systems by induction, and by direct connection through grounding wires. It was, in all ways, a profound disturbance to the planet.

The Carrington Event has been extensively studied as a prime example of the exceptional hazard posed to our modern technological world from powerful solar storms. Had this coronal mass ejection hit the Earth in the twenty-first century, it would have decimated satellites and their operations, and caused widespread power outages and problems across our global civilization. In 1921 a less potent but still very significant storm also hit the Earth. This event became known as the New York Railroad Storm after it sparked fires in a train-control center near Grand Central Station in New York City, when 1,000-volt spikes hit the system. During 1989 a series of smaller storms caused blackouts across North America, along with a number of satellite interruptions and even sensor problems onboard the space shuttle Discovery, which happened to be in orbit during one of the solar events. More recently, in February 2022, a comparatively mild event caused the premature failure and atmospheric reentry of forty Starlink internet satellites in low-Earth orbit.

But in the modern age we've so far avoided anything on the scale of 1859, or on the scale of even older historical events, which seem to have occurred roughly every five hundred years or so. Rather ominously, one of those events, in the year 775 CE, may have been *twenty* times more powerful than the Carrington Event.[4] Evidence for this storm comes from the abrupt spikes it induced in carbon-14 isotope abundances in the environment. It's hard to imagine the devastation that an event like that would wreak on today's technology-dependent world.

These historical solar storms represent the extremes of what we think that the Sun can dish out to the Earth. But the ordinary, day-by-day behavior of our nearest star has other profound effects across the entire solar system. The Sun's photosphere, like that of most stars, is just the foggy visible-light surface, where the density of its hot atmosphere of ionized gases quickly tapers off. But the star doesn't really end there. Just as Earth's atmosphere extends outward in an increasingly tenuous envelope, the Sun's very outermost atmosphere, which we call the corona, extends for millions of kilometers—about a tenth of the way to Mercury's orbit. The corona is so hot that the kinetic energy of its particle constituents can exceed the gravitational potential energy at the Sun's surface. Consequently, these particles have, on average, more than solar-escape velocity, and they become the vast flow of solar "wind" that carries on out to the boundary with interstellar space that the Voyager probes encountered. Because that wind is made of starstuff, it mostly consists of protons and electrons, with some helium nuclei, along with trace amounts of heavier elements like carbon and oxygen, or even iron and nickel. It blows outward at hundreds to thousands of kilometers per second.

That steady flow of charged material across the system affects planetary atmospheres and even erodes and weathers the surfaces of airless worlds, including the Moon and its intriguing accumulation of oxygen and water. The flow also pushes evaporating cometary material into the reflective streamers that sometimes visit our night skies. And as we saw with Voyager 1's encounter with interstellar space, the solar wind creates an enormous, distorted magnetospheric "bubble" around the solar system and plays a critical role in shielding us from other distant galactic cosmic radiation. For interplanetary exploration or even near-Earth travel, these opposing properties of the solar wind—both protecting and threatening—have to be carefully accounted for to ensure that equipment and delicate biology stay safe.

Like many other interplanetary phenomena, the idea of a solar wind existed before we had detected it. Carrington's observations in 1859 supported the notion that solar activity might somehow reach across space to affect the Earth, and by the early twentieth century, physicists were toying with the theory that the Sun should be continually shedding material across the entire solar system. But until the advent of space exploration, there was little real proof. In 1958 early orbital missions like Sputnik 2 and the US Explorer satellites carried radiation detectors that confirmed that the Earth's equatorial plane is encircled by inner and outer doughnut-shaped zones of swirling particles, all temporarily trapped in the planet's magnetic fields. In the outer zone, most particles have come from the solar wind. In the inner zone, there's a substantial number from high-energy cosmic radiation. Altogether, these regions extend from about 600 to 60,000 kilometers and are named the Van Allen radiation belts after the pioneering space scientist James Van Allen. In 1953 his studies had found tentative evidence for the existence of these belts by using small balloon-launched, high-altitude rockets (or "rockoons").[5] These zones are an orbital hazard to be avoided in the long term but also part of the Earth's magnetic diversion of solar and cosmic radiation that shields the planet.

More data on the solar wind came in 1959, when the Soviet Luna 1 Moon mission detected electrically charged particles flowing through the space around it and helped usher in what would become the modern field of heliophysics—the study of the physics of the Sun. Heliophysics gained even more impetus as humans ventured farther from the Earth and were exposed to interplanetary space. In retrospect, the Apollo missions were extraordinarily lucky when it came to solar storms. Right between the Apollo 16 and 17 flights, in August 1972, was another of the twentieth century's most powerful Earth-intercepting solar outbursts. This coronal mass ejection caused disruptions of global communications systems and satellite malfunctions, and it precipitated a measurable depletion of the northern ozone layer. Remarkably, it also caused the accidental detonation of dozens of magnetically triggered US Navy sea mines near North Vietnam.[6] The Moon was hit as well by this storm, and had an Apollo astronaut been walking on the lunar surface at that time, he would have almost certainly received a massive and fatal radiation dose. Perhaps more than any other event, this storm convinced space agencies that monitoring the Sun and pursuing the study of heliophysics wasn't just to satisfy scientific curiosity; it was going to be an absolutely essential part of space exploration.

Some Sun-observing missions had already begun flying by then, with a series of satellite observatories launched by the United States called the Orbiting Solar Observatory Program (OSO). In 1967 the OSO-3 satellite was placed into Earth orbit and began monitoring the Sun in ultraviolet, X-ray, and gamma-ray light, and capturing data on solar flares. By 1973, an X-ray telescope onboard the short-lived space station Skylab had made images of the Sun in visible light, ultraviolet light, and X-ray light, and the first true solar probes had been launched. These twin spacecraft, Helios-1 and Helios-2, were a joint German-US project and were placed into large heliocentric orbits.[7] These orbits were highly elliptical, so at their far point the probes were close to Earth's orbit, but at their near point they were only 30 percent of the distance from the Sun that the Earth is: very close to Mercury's perihelion distance. At a position this close to the Sun, the intensity of sunlight is eleven times greater than at the Earth. As was done for the later MESSENGER Mercury mission, the Helios spacecraft were carefully engineered to avoid overheating, with highly reflective conical solar panels and fused quartz mirrors coating the exterior of the craft. Without these strategies, the exterior parts of the satellites could have easily reached 370° Celsius (698° F).

The Helios probes also became, for a while, the fastest-moving spacecraft that humans had ever produced. As they swung through their perihelion point in the Sun's mammoth gravitational well, they hit an eye-watering 70 kilometers per second (253,000 kilometers per hour). Since that time, there have been more than forty international missions designed to study the Sun directly, as well as to study the interaction of the solar wind with the magnetosphere of Earth and with other worlds like Mars. Some, like the Solar and Heliophysics

Observatory (SOHO), have produced years and even decades of continuous imagery of the Sun's outer surfaces, with views in visible light as well as in the ultraviolet and X-ray parts of the electromagnetic spectrum, where the star's extraordinary textures, flares, and constantly changing surface are the most vivid. (See Figure 9.1.)[8] As with many if not most stars, the outermost parts of the Sun's plasma atmosphere experience convection—the upwelling and downwelling of warmer and cooler material just like in a boiling pot of water. To gaze at the Sun closely is to see this roiling stew of matter. SOHO was launched in 1995 as a joint mission between the United States and Europe, and it continues to operate today from its perch at the Earth-Sun L1 Lagrange Point—that place of orbital calm directly between us and the Sun.

An unexpected by-product of missions like SOHO has been the discovery of thousands of sun-grazing comets.[9] These ice-rich bodies flare into ultraviolet and X-ray brilliances as they plunge around, or sometimes into, the Sun. SOHO alone has spotted more than three thousand comets, many of which would have otherwise gone unnoticed, and has helped us better understand that ancient population.

There are currently some twenty-seven operating spacecraft orbiting the Sun that are devoted to monitoring its behavior and effects. The closest we've ever come to its surface is within 6 million kilometers (about 10 times the Sun's radius) in 2024, with the Parker Solar Probe. This heat-proofed spacecraft is capable of diving into the Sun's gravitational well at a blistering 692,000 kilometers per hour, or 0.064 percent of the speed of light—the highest speed ever attained by any interplanetary mission. It has also been the first mission to fly directly through the solar corona this close to the Sun, having a thick heat shield of painted white carbon composite and carbon foam that keeps the spacecraft at operational temperatures while the outside of the shield reaches more than 1,300 degrees Celsius. Even some parts of its solar-panel system are actively cooled by five liters of circulating distilled water. The Parker

Figure 9.1. A composite (negative) image from SOHO data showing the Sun's outer surface in extreme ultraviolet light (lower middle) and the streaming material of the corona, together with a coronal mass ejection (upper left) taking place from the Sun's surface in January 2002. The small black dot at the lower right corner of the image is the approximate size of the Earth, for comparison. (Credit: Courtesy of SOHO consortium. SOHO is a project of international cooperation between ESA and NASA.)

Solar Probe has to be highly autonomous in order to ensure that it constantly protects itself by correctly orienting its shield against the Sun. This autonomy is vital when signals take a round trip of sixteen minutes to and from the Earth.[10] And much like MESSENGER or Juno, the spacecraft uses a highly elliptical orbit to minimize the time that it's exposed to excessive radiation while it exploits Venus's gravitational pull to reorient and adjust its orbit at its aphelion points.

Being farther away from the Sun in the solar system does moderate the effect of solar storms, but only by degrees. In 2022 the MAVEN orbiter around Mars registered the impact of a coronal mass ejection that swept across the planet a few days after its eruption 230 million kilometers away. Even though Mars has no strong dipolar magnetic field (no north and south magnetic poles), enough of the solar particles were funneled into its thin atmosphere to create auroras across the planet. These included a daytime ultraviolet aurora when the solar particles collided with hydrogen in Mars's upper atmosphere. In another event, later in 2022, the density of the solar wind at Mars suddenly dropped, and MAVEN was able to sense that the Martian atmosphere on the dayside of the planet had swelled to four times its usual altitude, unburdened by the usually relentless pressure of solar material. The cause of this drop in solar wind pressure was a fast-moving pulse of solar wind that swept across slower regions and, like a large truck sweeping past you on a highway, created a temporary gap in the flow.

Farther away still, even Jupiter isn't immune to the Sun's effects. When coronal mass ejections intersect the gas giant's territory, they can squeeze its very substantial magnetosphere and shift the boundaries of that zone by a million kilometers. Like all things at Jupiter, even the auroras are supersized, and they can light up polar regions larger than Earth's total surface area with intense ultraviolet and X-ray light. Solar storms can also trigger ultraviolet auroras on Saturn and were first noticed when Pioneer 11 flew by in 1979.[11] In 2013 the Cassini mission experienced a solar storm sweeping across the Saturnian system that compressed Saturn's magnetosphere to within Titan's orbit, exposing the moon to the brunt of the storm's force and creating a shock wave across the top of its hazy atmosphere.

For airless bodies, such as our moon, the quiescent and active solar wind irradiates surfaces and implants ions and atoms. This is a very significant part of what is called space weathering: a phenomenon that also includes the impacts of micrometeorites, as well as the sputtering effects of particle radiation, that together can vaporize and erode exposed surfaces. Over time, space weathering causes many physical changes, including the formation of a distinct surface patina on rocks. Radiation also induces a mix of chemical changes, sometimes driving the formation of a dark reddish mess of organic compounds that coats many ice-rich asteroids and Kuiper belt objects. Consequently, space weathering is a factor to consider in the choice of materials used for interplanetary spaceflight, although detailed research on the long-term effects on spacecraft construction is at present quite limited because most deep-space probes don't return to be inspected.

Being able to predict all of the solar wind's behaviors and the occurrence of specific solar outbursts would be fantastic, but it is a work in progress. What we know of the physical mechanisms that govern the complicated movements and energetics of the Sun's magnetic fields and atmosphere still has its limits. However, there is one feature of the Sun's behavior that has been recognized for centuries, and that is called the eleven-year solar cycle.[12] Critically, this cycle seems to have effects on all space exploration, particularly exploration beyond the vicinity of the Earth.

Seen in visible light (when nicely filtered for delicate human eyes), the Sun's photosphere is mottled and textured, with its rising and falling convective cells of atmosphere. Sometimes, though, there are also dark-looking patches—sunspots—that persist for anywhere from hours to months. Sunspots were part of what intrigued Richard Carrington in the 1800s, and at that time some researchers suspected that these might be holes in a hot outer layer of the Sun's atmosphere offering glimpses of a cooler surface beneath. They were only partially correct in that interpretation, and in the early 1900s a newly discovered phenomenon of atomic physics, in which magnetic fields alter the spectral properties of atoms and ions, was used to show that these sunspots are not holes but actually correspond to stronger local magnetic fields. And these fields seem to block the usual convective turnover of hot atmosphere from below. That makes these patches cooler than their surroundings by as much as 2,000 degrees Celsius, and consequently—like the dimmed filament of an old-fashioned incandescent lightbulb—makes them emit far less light.

But the number and distribution of sunspots on the Sun's surface vary in time. Records in western Europe going back to the 1700s show that roughly every eleven years there is a peak in the average number of sunspots (and, conversely, every eleven years there is a minimum). That peak count can significantly exceed a hundred spots per month, and then over a few years it declines to episodes with effectively zero spots for months on end before the activity climbs back up. When the Sun has a peak number of sunspots, it usually also produces the most flares and coronal mass ejections. In other words, the eleven-year solar cycle is a gauge of the level of the Sun's storms and their effects across the solar system.

We still don't fully understand what drives this cycle. Similar kinds of cycles seem to exist on other stars, modulating their activity. For the Sun, a working hypothesis is that we're seeing the consequences of interactions among major magnetic structures, in this case a repeated game of one-upmanship between magnetic fields that align north–south and fields that align east–west. Or, more technically, this is an interplay between poloidal and toroidal fields generated by the rotation of the Sun (which spins at varying rates from equator to poles but on average once every twenty-seven Earth days) and the upwelling of material from its deeper layers. Consequently, the eleven-year cycle also seems to correlate with the overall magnetic-field strength and polarity of the Sun. That magnetic structure determines much of the quality of its solar-system–encompassing heliosphere—the teardrop bubble that usually keeps interstellar space at bay out to beyond the orbit of Pluto. In general, the heliosphere is actually divided in two by an unseen, undulating surface in the extended solar magnetic field. This invisible surface is a sheet of electrical current that follows the shape of an Archimedean spiral, forming a great twisted, rotating structure caused by the Sun's own rotation. This is called a Parker Spiral, after the solar physicist who first predicted it, and it also causes the jets of coronal matter and solar flares to spiral like a garden sprinkler, carrying billion-amp electrical currents out to the edges of the heliosphere in a vast electrical circuit.

The Sun has other, much-longer-acting, cycles that occur across thousands of years and whose effects are reflected in the abundance of radioactive carbon-14 in soils and fossils on the Earth. But the eleven-year cycle is perhaps the most important for space exploration on human timescales. During a peak in the cycle, the odds of dangerous solar storms rise significantly. But at the same time, increases in the Sun's activity and its overall magnetic field strength can inflate the heliosphere. That helps to shield all the major planets from galactic cosmic rays—the very-high-energy particles that can plow through meters of solid matter to cause damage to materials and biological cells. There's a catch, though: during the peak of the solar cycle, the overall polarity of the Sun's magnetic field also undergoes a reversal, exchanging its north and south like a flipped bar magnet. In practical terms, what this means is that although the strength of the Sun's field hits a maximum during the peak of its eleven-year activity cycle, there is also a time where it can drop and run through zero while it flips.

That field reversal is far from straightforward, with northern and southern hemispheres of the Sun changing fitfully and at different times, separated by months or even a year. As this takes place, sunspots of opposite magnetic polarity migrate across the surface of the Sun, gathering to the north and south. The summed effect is that both hemispheres will pass through their minimum field-strength level in these periods. This can disrupt the large-scale structure of the heliosphere and the spiraling sheets of current-carrying solar-wind particles that ripple through it. In some instances, this might weaken or shrink its shield-like properties against cosmic radiation. That's particularly important for human exploration to a place like Mars, where the minimum transit length from Earth in a Hohmann orbit is around six—highly exposed—months.

All of this scientific attention to the Sun's cycles has yielded other intriguing but worrisome insights about what's coming in the next decade or two. Over the past twenty-five years there seems to have been a steady decline in the peak polar magnetic-field strength of the Sun and its average activity level. In fact, there are indications that the Sun could be headed for a minimum in activity that hasn't been seen since a time from the mid-1600s to the early 1700s. This was an episode called the Maunder Minimum, after the English solar astronomer Edward Maunder, who studied it in the late 1800s. We don't really know what all the implications of this are. It has been suggested that the Maunder Minimum was connected to the "Little Ice Age," which produced some brutal winters across northern Europe and North America all across this period. But that may be largely coincidental.

What we do know, though, is that during this era the solar magnetic-field strength weakened, and this reduced the heliosphere's barrier against galactic cosmic-particle radiation. We find evidence of this in distinct increases in the amounts of radioactive carbon-14 and beryllium-10 isotopes produced on the Earth during the Maunder Minimum because of fast-moving cosmic particles hitting atomic nuclei in the planetary atmosphere and surface. If another long minimum causes a general increase in cosmic radiation across the solar system, this could have serious implications for deep-space exploration.

Irrespective of whether or not we're approaching another extended period of minimal solar activity, the radiation environment in interplanetary space is not to be trifled with. Research into the implications of radiation for space exploration has been carried out since the 1950s, along with efforts to mitigate exposure for humans.[13] That mitigation can involve limiting individual exposure or providing shielding and shelters, as with the case of orbital space stations. For interplanetary journeys, shielding is a major design consideration, not least because of the extra spacecraft mass it might entail. Luckily, some of the best shielding material turns out to have relatively low mass and is made of stuff we'd bring along with us anyway.

The physics of radiation shielding is complex, largely because of the different kinds of radiation that have to be dealt with. First, there is electromagnetic radiation, which can range from longer-wavelength visible light to ultraviolet light to X-rays and all the way to the shortest-wavelength gamma rays. Each of these is progressively more energetic, more penetrating, and more prone to causing damage if that energy is absorbed by a spacecraft or delicate biological cells. Separately, there is solar and cosmic-particle radiation, each of which consists mostly of high-speed charged particles such as electrons, protons, and other atomic nuclei. The cosmic particles streaming in from the rest of the galaxy through the heliosphere tend to carry the most kinetic energy. As they interact with planetary atmospheres or surfaces, they can generate more exotic, albeit short-lived, subatomic particle species such as muons, pions, positrons, and anti-protons or uncharged particles like neutrons, all capable of causing damage.

Blocking radiation coming from X-rays or gamma rays is generally easiest with dense materials with massive atomic nuclei, such as lead or steel, that simply intercept and absorb the high-energy photons. But as we've discussed earlier, high-energy particle radiation can be surprisingly insidious because those particles cause secondary effects. For example, speeding electrons—called beta particles—can be stopped by something like lead, but their deceleration can actually generate X-rays. Other high-energy cosmic particles can be stopped dead by massive nuclei, such as those found in a block of lead, but all that energy has to go somewhere, and it can produce a spray of new particles like neutrons that sail on through the shield largely unimpeded. This same kind of process occurs in places like the lunar surface or in Martian soil, and some scientific missions even exploit the phenomenon to detect the neutrons emerging from these surfaces in order to gain insight about their composition.

If, instead, high-energy cosmic particles like protons or helium nuclei (alpha particles) encounter light atomic nuclei, as found in materials rich in hydrogen, the results are rather different because of the way that our old friend momentum is exchanged and conserved. A particle like a proton that hits a hydrogen nucleus—also a proton— is likely to bounce off this equal mass, but it is also likely to hand off some of its momentum and energy to the other particle without causing a great cascade of new radiation. This is similar to how moderators work in nuclear reactors—they slow down fast neutrons but don't stop them all at once.

Consequently, an important component of shielding during interplanetary travel or on the surfaces of the Moon or Mars will be material filled with light, "bounceable" nuclei.[14] This makes water, with its hydrogen nuclei, a great option, as well as a number of plastic compounds like polyethylene that are full of light hydrogen and carbon nuclei. Other organic material can also work well, including human waste, which might be used to fill and replenish a shielding layer and then serve as an enduring punch line about the pleasures of space travel.

In principle, artificially generated magnetic or electric fields could be used to provide some shielding in space—perhaps diverting particles in the same way that the Earth's magnetic field creates a hull-like bubble around its local space. Some engineering concepts for this kind of shielding propose using sets of high-temperature superconducting electromagnets on extended arms. These electromagnets could finely shape a magnetic field so that it encloses a spacecraft but doesn't pass inside, where it might pose issues for electronics or astronauts.

The damage that radiation does to biology is one thing, but it's not a cakewalk for technology either, especially as robotic devices are given more and more sophisticated computing power and autonomy. High-end microprocessors are now so complex and so tightly packed—with over 100 million transistors per square millimeter (about the size of a pinhead)—that particle radiation tearing through these microscopic architectures can be like shooting an artillery shell through a collection of porcelain. The lattice structure of very precisely designed and formed crystalline semiconductor materials in chips can be permanently altered. Ionization tracks from radiation inside chips can either be a minor nuisance or irreversibly impair transistors' ability to operate properly. The sensitivity of some of these devices can be acute. In the 1970s it was discovered that the normal ceramic packaging material of certain dynamic-memory chips contained small amounts of mildly radioactive contaminants that caused endless "soft errors," where unwanted electrical charges would mess up the ones and zeros stored in the chips.[15] Even today, it is well understood that soft errors are a fact of life on the Earth, so chips and their firmware are designed to check for, and correct, errors induced by ordinary environmental radiation.

But in space, not only is radiation more prevalent, but the consequences of soft errors or more permanent damage can also be exceedingly problematic and hard to correct for. Since the earliest days of spaceflight, engineers have worked on ways to build radiation-hardened electronics, often exploiting the developments made in fabricating toughened military devices. There are a lot of different ways to do this. For instance, chips can be layered with a sapphire substrate (a form of aluminum oxide) that helps block radiation. The fundamental design of logic gates can be altered so that they are naturally better at resisting soft errors. Sometimes, just using physically bigger chips, with greater built-in redundancy in the number of transistors and logic gates, can mitigate the effects of radiation on function. All of these options tend to be expensive, though, because they involve heavy customization and smaller production volumes, which are of less commercial interest. One of the most successful radiation-hardened microprocessors is the RAD750, a device that has been used in more than two hundred spacecraft, including the James Webb Space Telescope. Despite its decades-old architecture, it costs around $200,000 per chip.

In some circumstances the only solution for avoiding radiation as much as possible is to place your most sensitive electronics in a totally shielded environment. This is an approach used in the case of Jupiter's intense radiation environment, which envelops the orbital zones of intriguing moons like Io and Europa. In these conditions, current spacecraft designs simply can't hold up forever against the barrage of energetic particles. A mission like the Juno orbiter, which has been in the Jovian system since 2016, carries a radiation vault that consists of one-centimeter-thick titanium sides that form a cubic meter of protected volume for the most delicate electronics. Without this vault, Juno's essential systems would have failed during its very first close pass of Jupiter.

The fact that we live in the midst of a constantly changing stellar weather system, in the ebb and flow of the winds and storms of the Sun, is something that most of us are blissfully unaware of. But it has been a major factor in steering the ways that we build our spacecraft and the ways that we protect astronauts, cosmonauts, and taikonauts. Space weather has profoundly influenced how we operate the Earth satellites and power systems that run most of today's civilization, to such an extent that attrition is routinely expected for low-altitude spacecraft as a result of solar activity. Driven by curiosity but also by these challenges, we've created a network of spacecraft and Earth-bound telescopes that monitor the Sun's behavior, and we've learned to be acutely mindful of the calendar dates for exploration. That can mean gauging where we sit in the eleven-year solar cycle or predicting the travel times for coronal mass ejections that can traverse the inner solar system in less than a couple of days but sometimes even take weeks.

When the Carrington Event took place in 1859, there was little by way of a technological world to feel the effects. But now we have to pay attention, for the sake of our Earth-bound infrastructure, and for vital services like GPS and internet connectivity that are supported by constellations of thousands of susceptible satellites. The same is true, and will be ever more important, for our machines and people scattered across the solar system, from the Moon to Mars and beyond. The more vehicles and habitats there are, and the more technology there is to support these things, the more the Sun's activity becomes a key part of the equation of function and survival.

Once again, there's a pattern here. As with the huge changes brought to our species by observing the Earth from space, yielding a capability that has revolutionized our quantitative knowledge about the world, mapping and monitoring the Sun's behavior and its influence are changing the nature of our actions and choices. This affects how we build and operate power grids and satellite systems for the Earth and how we fabricate microchips and organize human space exploration. The predictable unpredictability of stellar behavior is now a major element of the inflection point for life's evolution from planet-bound to interplanetary, and into a state of Dispersal.

What we have to think about in the final two chapters is whether we're ready to handle this and all of the other implications as we drive life's spread and influence onto new scales of space, diversity, and time, changing all the rules that have previously governed our choices.

— 10 —

THE COUNTRY OF
A BILLION SHIRES

I never dreamed that islands, about 50 or 60 miles apart, and most of them in sight of each other, formed of precisely the same rocks, placed under a quite similar climate, rising to a nearly equal height, would have been differently tenanted; but we shall soon see that this is the case.

—Charles Darwin,
The Voyage of the Beagle

For five weeks in 1835, the *Beagle* sailed around the great volcanic archipelago of the Galapagos Islands while its crew busily surveyed and studied the land, ocean, and life-forms that they found. In this curious place, a thousand kilometers from any continent, young Charles Darwin encountered puzzles that would stay with him for years to come. The abundant species of animals and plants distributed across the Galapagos's 127 islands and islets resisted any easy explanation by prevailing ideas about the nature of life—some of which assumed that "creation" had populated the world with a fixed set of organisms. Many islands appeared indistinguishable in terms of their climate and terrain, and were within sight of each other, yet the mix of species that inhabited each of these nubs of land could be quite different. There were distinct variations among animals like tortoises, mocking thrushes, and finches, and sometimes a species wasn't represented at all on a neighboring island that might be just a few dozen kilometers away.

Darwin speculated that factors such as winds and ocean currents might prevent life from dispersing and mingling more easily between islands. But many of the species were clearly similar to those on the American continent far to the east. In his diary, Darwin frets over the fact that the islands are not all commonly populated with identical species, nor are they each home to a unique set of organisms. Instead, the islands harbor a weird mixture of what are clearly very similar yet equally distinct and diverse organisms. It's little wonder that these observations would stick with him as you can almost feel his pain in trying to comprehend what was going on in the Galapagos.[1]

In retrospect, it's not surprising that these puzzles weren't readily solved. Reconstructing the story of biological evolution or the story of the solar system from present conditions is not easy. The same is true for predicting what their future stories might be. The theory of biological evolution that Darwin eventually formulated showed how tiny, barely noticeable, probabilistic opportunities and attritions play a role in sculpting species across timescales that stretch from mere moments to millions and billions of years, and across millions of variations. These subtleties are buried in layer upon layer of complex interactions—from the level of molecules to the level of entire populations of organisms—all contributing to our puzzlement about what we find swimming, flying, and walking around us today.

The solar system has its own plethora of subtle interactions and events that play key roles in determining its configurations and contents. Some are tucked into the unpredictable nonlinearities—the butterfly effects—of orbital dynamics and into the chance shaping of planets and moons by singular events like collisions. Countless tiny variations in structure and arrangement can propagate into momentous and even chaotic changes in the future. Sheer physical scale is also an important factor. Life on Earth may have run and erased from view a trillion microscopic experiments in variation, one on top of the other and obscuring what had happened earlier in this finite planetary environment.[2] By comparison, the solar system is so enormous and spread out that fewer of its experiments overwrite earlier ones, which are still out there waiting to be found somewhere among billions of objects. Equally, there is enough space for future experiments to proceed without interfering with one another, like islands separated by enough distance to seem like they're alone in the ocean.

There is perhaps no better way to examine this crazy experimental box of interplanetary space and time than to take a trip to the asteroids. There is a well-known architectural divide between these objects, marked by those that orbit well within the realm of the major planets and those that orbit beyond. That outer-solar-system population transitions to the Kuiper belt and farther: the region that contains billions, perhaps trillions, of ice-rich objects that we'd also readily label as asteroids. But an enormous population lies far closer to the Sun and within the zones of easiest exploration. These asteroids orbit the Sun anywhere from near Mercury out to the gas and ice giants. Like the trans-Neptunian objects, these bodies are staggeringly numerous and varied, and I think they offer vital clues to a part of the interplanetary future.

The most famous concentration of these objects is the asteroid belt, a torus-shaped orbital zone between the orbits of Mars and Jupiter that contains more than a million asteroids larger than a kilometer across and millions more that are smaller. Despite their abundance, added together they only amount to about 3 percent of the mass of the Moon or about 10 percent of the mass of Earth's crust. And about a third of all of that matter is contained in the very largest object, the 1,000-kilometer dwarf planet Ceres.[3] The total surface area of all the belt asteroids—the real estate of their rocky outsides—is projected to exceed 15 million square kilometers. This is 50 percent more than the nearly 10 million square kilometers of the continental United States. Compared to the huge volume of space that these millions of bodies occupy, though, the entire belt is—by human standards—nearly empty. When our spacecraft pass through this region, the closest they are likely to ever be to any substantial asteroid is a distance of about a million kilometers.

Despite being so sparse by terrestrial standards, the asteroid belt is richly structured. As with the Kuiper belt, where Neptune's mass tends to coerce orbits into groups related to mean-motion resonances, in the asteroid belt, Jupiter's powerful gravitational pull governs another set of mean-motion resonances. In this case the repeated perturbations from Jupiter disperse asteroids that come close to the lower-ratio resonances, sorting belt asteroids into major groups according to orbital period. The orbital ellipticities of belt asteroids are also varied, and orbital inclinations—the "tilt" of the orbits—range from zero degrees to more than twenty degrees. If you examine all of these properties, you'll also find great clusters, or families, with similar orbital conditions that suggest that these objects have a common ancestry. In fact, there are as many as thirty of these associations, with names like Eos, Vesta, Eunomia, and Flora, each with thousands of members. The existence of families of asteroids was something first recognized by the Japanese astronomer Kiyotsugu Hirayama in the 1920s, and they're thought to originate from the collisional breakup, or excavation, of larger objects. Fragments of all sizes orbit the Sun in great streamer-like clouds, each acting like a dynamic, orbital archipelago. Because of their history, family members are also likely to have the same material compositions.

Just as for the distant Kuiper belt objects, these asteroids of the "mid" solar system represent a smorgasbord of different histories and events. A few are about as primordial as you can get in an old planetary system. These primitive, often carbon-rich bodies are composed of material barely changed since condensing out of nebula clouds more than 4.5 billion years ago—an interstellar breeding ground that was laced with the remains of earlier generations of element-producing stars. Other asteroids were once heated to their melting points by fearsomely radioactive isotopes of iron or aluminum that were fresh and decaying when the solar system was still young. There are agglomerations as well, Frankenstein monsters stitched together by gravity out of fragments of different bodies and mixed with primordial dust and rock. There are also extremely dense, massive chunks that are rich in iron-nickel alloys and rare metals. These bodies likely once belonged to the cores of long-since-shattered protoplanets or may have formed close to the Sun, where an original rocky patina of carbonates and silicates would have been boiled away.

Exploring asteroids introduces a different set of challenges and technological needs than does most other planetary exploration. The earliest spacecraft encounters with these objects were largely coincidental, with probes like Pioneer 10 in the 1970s coming within several million kilometers of a couple of examples. The first genuinely close approach (by cosmic standards) to a regular asteroid was also opportunistic, when the Galileo probe flew by the asteroid Gaspra in 1991 while heading to Jupiter.[4] The Gaspra encounter was at a distance of just 5,000 kilometers, and Galileo returned fifty-seven images showing a 20-kilometer-long potato-like body textured with craters and large, flat areas.

Gaspra has a particularly interesting backstory, at least for humans. These flat regions on its surface may correspond to where the asteroid broke off from a larger parent body (probably as a result of a collision), and it's thought that Gaspra actually belongs to the asteroid family called Flora. This family, named after its largest member, which is about 140 kilometers in size, consists of about thirteen thousand "stony" asteroids that are rich in silicon dioxide and have very similar orbital parameters. It's likely that most of these objects came from an original protoplanetary parent that was violently fragmented in an encounter with another body. Data from Galileo's flyby of Gaspra suggest that this collisional breakup was only 200 million years ago, and there is a possibility that another one of the Flora family members was the asteroid that hit the Yucatan Peninsula 65 million years ago, contributing to the mass extinction of species like the nonavian dinosaurs.

Subsequent missions have been specifically designed to visit asteroids. For example, hefty Ceres was the very first asteroid discovered, in 1801, and was finally visited in 2015 by NASA's spacecraft Dawn—using an ion-drive propulsion technology that we'll shortly see as key for the future of exploration. This was the first-ever mission conceived and designed at the outset to enter into orbit around one body circling the Sun and then to move on to orbit another. Dawn had arrived at and orbited the asteroid Vesta in 2011 before heading 70 million kilometers farther from the Sun to Ceres, which the probe's derelict remains continue to orbit to this day. Another multi-destination mission is NASA's extraordinary Lucy spacecraft, launched in 2021 and that I've mentioned before. Lucy is expected to take twelve years to visit at least six separate asteroids, including some in the L4 "Greek" and L5 "Trojan" groups of Lagrange point orbits leading and trailing Jupiter.

Historically, this type of space exploration was really kick-started by a desire to visit the objects we call comets—long puzzled over by earthbound human eyes. These days, though, cometary bodies are seen as less of a mysterious or distinct population and more as objects that just belong to a volatile-rich, or "wet," class of asteroids. Their formation history in the outer solar system has left them containing a lot of water and other frozen compounds, all of which can boil off if they approach the Sun, sometimes producing the classic cometary tail. There have been a dozen missions undertaken to comets, including the first audacious attempt in 1985, which used a "recycled" heliophysics spacecraft from 1978 that was originally called the International Sun-Earth Explorer-3. Coincidentally, this had been the first-ever mission to sit at the Sun-Earth L1 Lagrange point. Rebranded in 1983 as the International Cometary Explorer (ICE), it was directed away from its orbital perch, using five gravity assists from the Moon in order to intercept the plasma tail of the comet Giocabini-Zinner. We quickly learned that exploring comets in the Sun's vicinity brings some new challenges. The spacecraft environment can be complicated by the fine dusty material and gas that cometary bodies often shed. This was something that the European Rosetta mission also encountered in 2015, when it visited the striking dumbbell-shaped cometary nucleus of 67P/Churyumov-Gerasimenko and found its navigational star trackers confused by the comet's dust cloud and that same dust causing drag on the spacecraft's 32-meter-long solar arrays.[5]

Over the years, the Japanese space agency (JAXA) has built a special track record at comet and asteroid exploration that also illustrates the peculiar and special challenges of this type of activity. This story began in 1986, when the 15-kilometer, icy chunk of ancient material called Halley's Comet was making its thirtieth recorded perihelion pass of the Sun, coming almost as close as Mercury's orbit. In response, an armada of international spacecraft set out to intercept the famous cometary nucleus and its equally famous tail of glowing, flowing material.

That armada included the Soviet Vega 1 and 2 probes, after they had dropped their payloads at Venus, as well as a new European mission called Giotto and the ICE spacecraft (repurposed yet again). Alongside these were not just one but two probes sent by Japan. The Japanese spacecraft were stout cylindrical machines named Sakigake (or "pioneer") and Suisei ("comet") and were almost identical in construction, but with each carrying a different scientific payload.[6] Sakigake was Japan's first-ever interplanetary mission and had been launched in 1985, a few months ahead of Suisei, in order to iron out any issues with the launch system and the spacecraft function. In the end, it was Suisei that swung by the sunward side of Halley's nucleus to gather data, helping to pin down the amount of water boiling off from the comet's surface. Meanwhile, Sakigake provided reference measurements to support the other probes in the international armada as they made measurements even deeper inside the comet's shroud of gas and dust.

As modest as these accomplishments might sound, they established Japan as the only other nation at that time, after the United States and the Soviet Union, capable of interplanetary exploration, and they were doubly impressive because of the time-critical nature of the Halley encounter. The Japanese space agency has continued to send increasingly elaborate missions to asteroids ever since. In 2005 the Hayabusa (or "peregrine") mission rendezvoused with the asteroid known as 25143 Itokawa (named for the Japanese engineer Hideo Itokawa, who had played key roles in developing rockets for the country's space effort in the 1950s).[7] In doing so, the spacecraft made the first-ever—albeit rather tiny—collection of samples directly from a dry asteroid, samples that were returned to the Earth in 2010. To get to Itokawa, which orbits the Sun at between 95 percent and 170 percent of Earth's distance, the Hayabusa spacecraft used four gently thrusting ion engines almost continuously over a period of two years. The asteroid itself was quite a revelation, for it turned out to be a bizarre double-lobed, peanut-shaped object about 500 meters long. (See Figure 10.1.) Not only did Itokawa appear to consist of the merger of two smaller asteroids; those bodies were not solid rock at all but were literally "rubble piles" made of loosely associated boulders, pebbles, and dust.

Figure 10.1. Asteroid Itokawa as seen by the Hayabusa spacecraft in 2005. Its length is approximately 540 meters. The smooth areas are known as "dust ponds," where finer particles accumulate in depressions in the larger structure. (Credit: ISAS/JAXA.)

Later missions, including Japan's Hayabusa 2 and the United States' OSIRIS-REx, would encounter more of these rubble-pile objects, formed from the shattered and re-coagulated pieces of other bodies and primitive materials.[8] When OSIRIS-REx descended to the surface of asteroid Bennu in 2020 to retrieve samples to bring back to Earth, the spacecraft plunged almost half a meter into the asteroid. It was like dropping into a swamp, except that this swamp was made of small dry pebbles and dust.

The pebbly, dusty, rubbly nature of these asteroids and, by inference, possibly millions of other small bodies in the solar system was not entirely unexpected. When earlier missions had encountered asteroids, it was clear that these objects sometimes had a lower density than was possible if they were a single, solid, monolithic lump of material. This is consistent with our ideas about the formation of these objects in the solar system, where their history contains an endless litany of gravitational building and unbuilding. Even Mars's moon Phobos may be a porous rubble pile lurking inside a thin skin of dusty material.

For a species used to strong planetary gravity and solid rock, these are very disconcerting properties. You could probably "swim" through some of these asteroids, with a little care and perseverance. They are more akin to globules of cosmic quicksand than to any solids we're used to thinking about. But they're also rich in interesting compounds and could be a bounty of resources. Their loosely bound contents certainly lend themselves to being "harvested," either by scooping material up or by blowing jets of gas to waft the pieces into a container.

Orbiting these gnarly objects, with their weak and unsymmetrical gravitational fields, is a delicate and attention-demanding business. You can't just place a spacecraft into an orbit and step away; maintaining a configuration requires constant monitoring and adjustment. In the field of asteroid exploration, there is some pride taken in just how tight an orbit can be achieved. When the European Rosetta mission surveyed the cometary nucleus of 67P/Churyumov-Gerasimenko in 2014, it was able to find a moderately stable orbit just a little over 7 kilometers from the extremely nonspherical dumbbell-shaped nucleus. When the OSIRIS-REx mission established a surveying orbit around the asteroid Bennu in 2018, it was placed with delicate care into a trajectory just 1.75 kilometers around the half-kilometer-diameter object. In that surveying orbit, the spacecraft circled every 62 minutes, largely held to this path by Bennu's gravitational pull, a force that is just five millionths of that at Earth's surface.

To find that stable orbit took some serious computer modeling, using data collected as OSIRIS-REx homed in on Bennu and could sense the faint pulls of its uneven mass. But even in this orbit, the tiniest forces from solar-radiation pressure and the thermal radiation from Bennu itself meant that the spacecraft had to regularly correct, or trim, itself using its thrusters. Because of the range of demands during its mission, OSIRIS-REx carried a total of twelve hydrazine engines and sixteen thrusters with differing thrust characteristics.

Finding a safe harbor in orbit around these small bodies is a special task. It's very different from but at least as challenging as nailing an orbit around a gas giant like Jupiter or Saturn. There are a lot of tentative approaches before rendezvous, a lot of cautious experimentation, and a need for the gentlest of touches on thrusters.

However, there are other, less delicate, ways to study asteroids and comets. In 2005 the United States' Deep Impact spacecraft placed a meter-sized package directly in the path of a cometary nucleus called Tempel 1. The package was laden with scientific instruments that were bolted to a 100-kilogram structure made of pure copper. This copper was dubbed the "Cratering Mass," and its purpose was to bulk up the probe so that when it and the comet collided at a combined velocity of over 10 kilometers per second, it would be like setting off five tons of TNT. The point of this violent exercise was to excavate and study the innards of Tempel 1, and copper was not expected to exist in any abundance in the comet, so it wouldn't contaminate measurements of the body's composition. The impactor package was also "smart," and it used thrusters to autonomously adjust its position to make sure that it was on target for a collision. In the meantime, the main Deep Impact spacecraft would observe the whole process from a safe distance and swoop past a few minutes after the event, barely 500 kilometers away.

Remarkably, this all worked to plan, and the impactor carved a 150-meter-wide crater in Tempel 1, ejecting a great plume of material that shone brightly in the sunlight. Data revealed that the comet's nucleus was far dustier and drier than expected, as well as extremely porous. In fact, Tempel 1 seems to have a structure more akin to a very filthy snow drift during the first warm days of spring. The size of the crater was confirmed six years later, in 2011, when another mission, called Stardust, was redirected to fly by Tempel 1 and snap images of the unobscured impact site.

Deep Impact was a gloriously audacious project. For it to succeed, that smart, autonomous targeting system—called AutoNav—had to handle its own impending fate at a distance of over 130 million kilometers from the Earth, where signals were taking a little more than seven minutes each way.[9] As the designers were happy to point out, the doomed AutoNav had only a few moments to look into unknown terrain, figure out where Tempel 1's sunlight regions were, and compute how to position itself to selflessly nail its target at 36,000 kilometers per hour.

Deep Impact did something else, too; it marked a moment where we can be certain that we modified the orbital motion of a natural celestial body. When the Tempel 1 cometary nucleus collided with the impactor, its orbital motion was estimated to have changed by 0.0001 millimeters per second and its orbital period around the Sun by 1 second. That's not much, but it was far larger than any direct effect we'd previously had on any other celestial body.

In 2022 our species carried out a deliberate act of cosmic reconfiguration by colliding the 600-kilogram Double Asteroid Redirection Test (DART) spacecraft head-on into a 200-meter-size object called Dimorphos.[10] This small asteroid, or "moonlet," orbits a larger object called Didymos with a leisurely motion of some 0.18 meters per second, making it relatively easy to target and then measure changes to its movements. Dimorphos's status as a moonlet, a part of a binary asteroid, also provides an extremely conservative safety net to reduce the chances that DART might cause either asteroid to be—in some unforeseen circumstance—diverted to an orbit that would pose a hazard to the Earth.

When the DART impactor plowed into the rubbly, loose composition of Dimorphos at 24,000 kilometers per hour, it changed the asteroid's velocity by a Delta-v of 3 millimeters per second, decreasing its orbital period around Didymos by an astonishing 33 minutes out of a 12-hour orbit. The collision also created a 10,000-kilometer plume of dust and boulders that followed the asteroid for several months before dispersing. It's thought that DART's impact likely also turned Dimorphos into a chaotically tumbling object. In 2026 a European mission called Hera has plans to reach this pair of asteroids and do a full check-up of the system.

The success of DART may seem like a relatively minor event in the grander scheme of the cosmos. It was, after all, mainly designed to be a proof of concept for a means to divert small asteroids that pose a hazard for the Earth. But it has implications that are far reaching. Subsequent research has examined what will happen in the future to the boulder-sized pieces that DART ejected from Dimorphos. Applying the mathematical tools of gravitational dynamics, computer models calculate the boulders' pathways and show that in about six thousand years, some of them may intersect Mars in its orbit.[11] If that happens, these objects will streak into the Martian skies as meteorites, hitting the surface and digging a new set of craters. This means that the actions of living organisms, millennia before, will not only modify the substance and orbit of a single asteroid but will also create a ripple effect that will ultimately modify the surface of another world.

DART paves the way for our species to lower its own existential risk due to phenomena like asteroid impacts, but it also marks the point where our species learns to deliberately and significantly reshape the form and natural placement of matter in the solar system. That is an inflection point for all life because, more than anything else, the billions of asteroids in the solar system represent raw, accessible, malleable ingredients. They resemble the once barren but rich volcanic archipelago of the Galapagos on the Earth. As such, asteroids offer a window of insight about where space exploration may take us and a way to begin to fully address the most fundamental questions about what happens to life when a species breaks through the barriers of existence on a single planet.

Predicting the future is, of course, a hazardous business. But as with the contents of the solar system, we can try to find the boundary conditions—the limits of the future. To do that, we can take a deliberately heavy-handed, expansionist look at the possibilities of sculpting the solar system in ways that we've barely begun to imagine. I should, and you should, be very clear about this: not every human culture would think this way, and it's not yet obvious whether there is any practical, ethical, or moral rationale for doing some of the things being imagined. There is a counterpoint to that hesitancy, though, which is that evolution may inevitably push us up to the boundaries, regardless of what we think. Later, I will describe a more palatable version of all this future extrapolation, one that is even desirable, for it embraces what we truly are as a species and what we might become without ignoring our tumultuous histories or contemporary problems.

The first point of reference to orient us, as we look for these future boundaries, is the present size of the Earth's "technosphere," sometimes also called the "noösphere" (the sphere of the mind). That's all the buildings, bridges, roads, dams, tunnels, walls, refined metals, and intricately engineered substances or machines that humans have forged from the ancient matter of the Earth. Present estimates indicate that all of this totals up to around 30 trillion tons of stuff.[12] If spread out evenly, that would amount to approximately 50 kilograms of engineered substances for every square meter of the surface of the planet. Remarkably, the mass of the technosphere now exceeds the mass of the biosphere—the sum of all living things on the planet.

Yet as absurdly big as the technosphere is, it still represents only 0.00015 percent of the mass of Earth's crust. Or to put this another way: in principle, we have a very long way to go before we are even remotely close to using up the fundamental matter resources of the planet, especially if we disregard the needs of the biosphere. Equally, this suggests that the asteroid belt by itself, if it were mined and engineered, might be converted into many, many technospheres the size of the one that exists on Earth today. That's interesting because although a collection of matter like the asteroid belt is spread across a vast region by human standards, a virtue of that spread is that this material is also extremely accessible in cosmic terms. Arguably, it is a lot more accessible than the metals and rocks that are buried on top of each other throughout the Earth's crust and held onto tightly by the planet's considerable gravitational pull.

The audacious back-of-the-envelope calculations you're about to see (the kind that I readily admit physicists are all too eager to fall prey to) take this premise and ask what the limits are for the future— the boundary conditions for a species. Specifically, if the elemental composition of the asteroid belt is mostly similar to the Earth's crust, how many more of Earth's technospheres could be built from this material? The answer is that a species could use the asteroid belt to construct about eighty thousand technosphere-scale systems without ever touching another planet.

Of course, you can't build things without using energy, so how much energy would the construction of eighty thousand technospheres require? We can gauge the rough scale of this by considering the total estimated energy needed for humanity to build and maintain its current technosphere here on Earth.[13] Summed over the past twelve thousand years, this is thought to amount to about 37 zettajoules (10^{21}), which probably doesn't mean very much to you. For a slightly more everyday comparison, if we were to produce that amount of energy in a single year, we'd need to generate about 1.1 million gigawatts of power nonstop during that year. That would be about 150 times more electrical-power-generation capacity than we presently have across the planet.

If each asteroid-derived technosphere calls for its own 37 zettajoules of energy in order to be constructed and run—assuming that it would be accomplished by growing the technospheres over centuries—that would require a colossal power budget. Yet on the scale of a planetary system and the solar-power budget, it's actually not so bad. To create all eighty thousand technospheres would require 550 years' worth of the total solar power hitting the Earth's upper atmosphere, or a mere 8 seconds' worth of the total power output of the Sun.[14]

In other words, as absurd as this may all seem, there is enough matter in millions of small bodies and enough raw energy in the solar system to—in principle—construct and operate tens of thousands of technospheres, each the size of what exists on Earth today. If those technospheres were there to support living things and new biospheres, that wouldn't be a problem either, with matter enough to go around if we were careful: all without digging into another planet.

Let's pause to digest all of this for a moment. For a spacefaring species, as we are becoming more of at an ever-accelerating rate, the *possibility* of creating new places or infrastructures elsewhere is entirely supported by the amount of available matter and energy in the solar system. Plenty of that available matter is not tied up in major planets (where it is deeply buried and sunk in gravity wells) but is instead already formed into accessible chunks. Some, perhaps most, is present as rubbly, pebbly swarms of readily processable matter.

Of course, constructing habitats and technospheres is meaningless if there aren't people and other living things to occupy them. Here, too, we have to think a little differently and ask how much matter in the solar system is suitable and available to "grow" humans and other living things. In 2005 the researcher Michael Mautner published an intriguing study of the availability of key biological elements like phosphorus and nitrogen—vital for life but not always very abundant—as well as the availability of carbon and other important components in the rocky asteroids and other small bodies of the solar system.[15]

Mautner's calculations indicate that if you could harvest all of those objects in their hundreds of millions, there are enough suitable elements to make at least six thousand biospheres the size of Earth's current biosphere (ignoring the matter required for technospheres). Given that one biosphere currently supports some eight billion humans, the implication is that the rubbly, dusty elements of the solar system could be converted into enough living material to create and sustain somewhere around fifty trillion humans in fully fledged biospheres. If you up the ante by including the matter available in the major planets or perhaps just sacrifice the matter in Uranus and Neptune (because aren't they just a little boring?), then all of these figures become even more colossal, with the possibility of millions of trillions of humans and tens of billions of biospheres and techno-spheres, all readily sustained by some portion of the hundred trillion trillion watts of the Sun's power output.

To state all of this slightly differently, in the grander cosmic scheme of things, even within just our one solar system, the extent of our present circumstances is, well, kind of pitiful. Life may have spent four billion years growing and evolving to cover the Earth like some eagerly grasping lichen on a newly emerged volcanic island in the Galapagos, but that singular emergence has, until now, also left us as a small and quite parochial oddity. That condition is clearly not the limit, though, and the asteroids are a clue to the bounds of what is possible, even if the details of the actual implementation of all this construction remain laughably vague. The real challenge here is for us to learn to imagine a solar system so differently configured, with life and all of its supporting structures on such a scale.

How could life begin to modify the organization of matter like this? If you recall the tyrannical rocket equation and the constraints of Delta-v on orbital maneuvering, it might seem unlikely that such wholesale repurposing could ever happen. Visiting and harvesting every asteroid, tweaking their orbits, and assembling infrastructures would surely cost so much propellant—even if derived from the same asteroids—that it would be self-defeating. However, we already know of ways to fix this, with many missions to asteroids and com-ets already deploying technologies and orbital strategies well suited to long, patient trips and operations in the gentle gravitational fields that these objects produce.

The ion thrusters that Goddard and Tsiolkovsky invented in the early twentieth century are a key technology.[16] The very first in-space tests of this form of propulsion were carried out by the Soviet Union and the United States in the late 1960s and early 1970s. Small thrusters were developed that, in essence, applied a voltage to a metallic grid that would pull and accelerate ions of propellant to speeds approaching 50 kilometers per second. These speeding ions would pass through the grid to provide a propulsive jet. In all current ion-propulsion systems, the instantaneous thrust—the force generated—is usually far, far smaller than that of a chemical rocket, but the velocity of the expelled propellant is far higher (yielding a very high specific impulse, to use the technical terminology).

A run-of-the-mill modern ion drive, used for interplanetary exploration, might generate only a hundred millinewtons of thrust, or about the same force as you feel when you rest a sheet of paper on your hand. But it does that by expelling only a few milligrams of inert propellant, like xenon gas, per second. This means that with a few hundred kilograms in a tank, you can accelerate continuously for years on end and actually reach speeds of at least as high as 10 kilometers per second, all aided by Oberth's magical effect, which makes it more and more efficient to pile on kinetic energy as time goes by.

Because of these characteristics (and the fact that an ion thruster can be as small as just 10 centimeters in length), low-thrust, basic ion propulsion is already very common on Earth satellites to help them keep on the right orbit. Geosynchronous satellites are often equipped this way, for example. But ion thrusters, or drives, for interplanetary voyages have taken a little longer to come into their own. A pivotal technology-demonstration mission was NASA's Deep Space 1 in 1998. This used an experimental ion drive to visit both an asteroid and a cometary nucleus. The mission also tested one of the first intelligent, autonomous systems to self-diagnose problems and to make decisions about the spacecraft's activities.

Today, major scientific missions like the United States' Dawn mission to the asteroids Vesta and Ceres have used ion drives. So does the European Space Agency's BepiColombo mission, which will arrive at Mercury in late 2025 after an elaborate seven-year orbital journey with no fewer than nine planetary flybys and gravity assists after its launch in 2018. BepiColombo's four grid-based, solar-powered ion engines exhale a beautiful blue-glowing jet of xenon ions to produce a few hundred millinewtons of thrust. Together with the mission's multiple planetary assists, these engines help produce several kilometers per second of Delta-v, allowing the spacecraft to drop deeper into the Sun's gravity well and to catch Mercury as it scoots along in its orbit at 47 kilometers per second.

Ion propulsion is also particularly well suited for capturing the energy of sunlight to generate the electricity needed to accelerate its chemically inert propellants, and there are a wide range of possibilities for designing these engines. Aside from the grid method, another approach is to use a Hall-effect thrust, which cleverly does away with a physical grid to carry the accelerating voltage. Instead, the Hall-effect thrust uses magnetic fields and a negatively charged cloud of electrons (a plasma) to ionize, accelerate, and then electrically neutralize the propellant atoms. Without a fragile grid sitting in the exhaust plume, Hall thrusters can potentially last much longer and can be scaled up to produce much more thrust. They can also be made small and have been used for decades in hundreds of Earth-orbiting satellites. More recently, that number has grown to thousands, with the ever-expanding constellations of low-orbit satellites like the Starlink internet system using ion thrusters to help these compact devices maintain their altitudes. As I write this, NASA's Psyche mission is headed for an asteroid rendezvous with its namesake in 2029, using the first Hall-effect drive for interplanetary exploration. That drive is a four-and-a-half-kilowatt–rated brute powered by the spacecraft's 75-square-meter solar panels: a full solar-electric propulsion system.

The scalability and efficiency of ion drives is what a species can use to begin transporting and reconfiguring the material of the solar system. Chemical propulsion still has a few tricks up its sleeves, too, with new ways being discovered to squeeze ever more efficiency out of its ingredients. Powerful computer modeling can dissect the subtle but complicated properties of swirling and combusting propellants to improve efficiency. For instance, new "rotating detonation" engine prototypes coerce fuels and oxidizers into a self-sustaining cycle that races around a tubular chamber in an endless supersonic combustion. This type of engine appears capable of producing a mammoth 25 percent efficiency gain in converting combustion into forward propulsion. All this was made possible by advances in computation, materials science, and fabrication techniques like 3D printing.

Other key changes have been gestating across the past two decades in rocket-launch technology. A commercial space industry has existed for pretty much the entire history of modern rocketry, but companies are now churning out new launch systems and rockets at an extraordinary pace. For example, in 2024 the SpaceX company reached a milestone when it manufactured a second-stage launch section (the upper stage for its Falcon 9 rockets) every 2.5 days. It may not yet be mass production, but it's edging awfully close to that.

Exponential growth in rocket launches has gone hand in hand with rapidly decreasing economic costs, driven by reusable stages.[17] In the 1960s, it cost a space agency like NASA about $100,000 per kilogram of launch mass (in dollars adjusted to today). By the time the Saturn V rockets were in use in 1967, those large launchers were yielding a cost of some $5,400 a kilogram. This was particularly efficient, though, because through the intervening decades until 2010, costs otherwise hovered around $15,000 to $30,000 per kilogram for medium- and light-capacity launch systems. Then SpaceX, with its reusable launchers, managed to push costs down again to $1,500 to $2,500 per kilogram, and other private launch companies like Blue

Origin are targeting close to the $1,000-per-kilogram mark in the near future.

That's a very big change, and independent projections of launch costs are betting that they reach around $100 per kilogram ten or fifteen years from now, especially if projects like the SpaceX Starship launcher (with a capability to place over a hundred tons into Earth's orbit) succeed.[18] The most ambitious of these private launch efforts are targeting a cost level of just $10 per kilogram to get into Earth's orbit (mostly in the cost of fuel). To put all of these figures in perspective: today you might pay $1,000 for a long international flight with an airline that moves your 70-kilogram average body mass a few thousand miles at a cost of $14 per kilogram.

There's little dispute that this steady reduction in launch costs is one of the biggest changes to space exploration since the 1960s. This has simply not happened before. Behind it are, for the most part, hopes to make money by speculating on the growth of the space economy. A lot of that economy is driven by an ever-increasing need for remote sensing data on the Earth. Those data have become a critical support for agriculture and governance in the face of rapidly changing environments, and they feed a seemingly unquenchable thirst for communications and dataflow. Estimates suggest that by the mid-2030s, these areas of the space industry will represent a $1.8 trillion annual economy.[19]

There is also an emerging commercial interest in the material wealth of asteroids. A 200-kilometer metal-rich object like the asteroid Psyche may contain elements such as iron, nickel, copper, gold, and platinum in quantities amounting to hundreds of trillions of tons. You can find newspaper headlines that gleefully announce the market value of these elements as being in the thousands of *quadrillions* of US dollars. But there are issues with this enthusiastic valuation. For one thing, the colossal expense that would currently be incurred to try to develop the means to harvest, refine, and return precious metals to the Earth will offset the value of those resources. That value itself is also a product of a terrestrial marketplace, where the very scarcity of these compounds is important. Instead, where the economics of asteroid materials starts to make real sense is in a market that doesn't quite exist yet, and that is the market of interplanetary space itself.

In the past century, we've watched humans place hesitant feet on the Moon and encircle the Earth with an orbital, robotic "exo-technosphere." Another, diluted, exo-technosphere exists even farther afield in the machines populating Mars and cruising through interplanetary space to park themselves around other planets or to speed ever onward. These hint at an emerging, evolving interplanetary infrastructure that is linked to our species' intent to return to the Moon and to commercial launch and exploration systems. That infrastructure will need many resources. It might be easier to supply the Moon or Mars with elements harvested from asteroids than any kind of indigenous sources. But what is the endgame, the natural trajectory for all of this? If the solar system offers raw materials and energy to forge innumerable new technospheres and biospheres, what would those look like?

We've already seen how a world like Mars, as alluring as it is, has innumerable issues as a focal point for life's extension away from the Earth. Instead of seeding technospheres and biospheres on planetary surfaces, there could be enormous advantages to being positioned out in interplanetary space. You can find or place your habitats or machinery at just the right distance from the Sun—near enough for the opportunity to gather plentiful raw energy from sunlight but flexible enough to support a biological system that needs to dodge solar storms and cosmic radiation. You can select places and orbits that are more protected from other asteroids and debris but are also situated so that resources are easier to reach and to harvest. These choices could also easily accommodate our sense of preservation and respect for the natural environment. In other words, you can extend the zones of easy exploration to incorporate zones of easy growth and evolution—like picking the choicest and least damaging options in the Galapagos archipelago.

Customized, adaptable structures in interplanetary space also provide unique options for mitigating the deleterious effects of low gravity. Anyone's who's ever watched Stanley Kubrick's 1968 film *2001*, based on Arthur C. Clarke's stories, or modern fare like *The Expanse* will have seen technically accurate representations of people walking or jogging around inside large space-based environments that spin about an axis. This spinning motion generates centripetal force—effectively pulling objects toward the axis in a circular motion as their frictional attachment to the structure coerces them to spin with it. To a human, that force is perceived as an outward push "down" toward the rim of the structure, a sensation we call centrifugal force. Physiologically, this feels like gravity on a planetary surface, or like having weight.

Strictly speaking, this is not an exact replacement for standing on the surface of a massive planet, for a couple of reasons. What Albert Einstein demonstrated in his General Theory of Relativity is that un-propelled (un-accelerated) objects must move along the shortest paths through the curved space and time produced by any mass. These paths are called geodesics, and some geodesics yield orbits if an object has the right velocity. What we previously learned to call "the force of gravity" is an approximation used to explain our motion in this curved space and time. On a planet you feel weight because the material forces between your body and the ground are preventing you from moving along a straight geodesic toward the planet's center. Uniform *linear* acceleration—the kind you get after you push the gas pedal in a car or shoot off in a rocket—is what's most directly equivalent to this experience.

The acceleration experienced in a spinning structure is not quite the same, but it is good enough to fool our physiological responses and to be our artificial gravity. In other ways, though, the practical implementation of a spinning environment has noticeable and sometimes disconcerting features. For example, imagine you're in a spinning hoop of a space station that rotates about an axis like a giant bicycle wheel. The centripetal force of that motion can keep your feet on the inner rim, with your head pointing toward the axis, and you feel weight. But if you're standing, then your head naturally ends up being a bit closer to that central axis, so it experiences a slightly different force because it's moving around a smaller circular path. That's not the same as when you stand on the surface of the Earth, where the difference in gravitational acceleration between your feet and head is absolutely tiny (it is, in fact, different by less than 1 part in 150,000), nor is it the same as in purely linear acceleration, where every part of you is accelerating at precisely the same rate. In a spinning structure, your sense of balance may feel a little off as your inner ear tries to deal with the situation. It gets even trickier if you bend down to pick something up because now you're moving your head and body through different amounts of circular motion and changing your circular momentum. Because momentum likes to be conserved, you'll feel a corrective force across your body as you bend over or stand up.

Or if you try to pour yourself a glass of water, the stream of liquid may miss the glass because it won't fall where you think it should. In the same way, if you climb a ladder to move "up" closer to the axis of the spinning environment, you will feel more of the force you felt when bending over—the Coriolis force—pushing you from the side as your momentum has to adjust to a smaller radius of motion. Things can get even weirder in a rotating structure if you try running in the direction of rotation, or against it. Jog along with the sense of rotation, and you'll feel heavier. Run against the sense of rotation, and if you're speedy enough you can find yourself floating away from the "floor" until you drift back into it or bump into the central axis.

These aren't the only challenges. If a structure is the size of today's space stations, it has to spin pretty quickly to generate a centripetal force equivalent to what we experience on the surface of the Earth. Imagine, for example, that you could take a 10-meter section of the ISS and tumble it to generate artificial gravity at either end. To match the acceleration felt on Earth's surface would call for the ends to have a rotational speed of about 7 meters per second, but that means the section would rotate about 13 times per minute, or 13 RPM. Experiments on the Earth with human subjects indicate that anything more than about 2 RPM messes with our inner ears at a level that is hard to bear and makes all of the weird effects we've just described very noticeable.

The solution to most of these problems is to go big. If the rotating structure was 200 meters across, for example, then the RPM needed to generate Earth-equivalent forces would be less than 3. Larger-sized structures have another major advantage, in that their mass is greater compared to all the stuff that might need to move around inside of them, like restless humans. If you've ever tried to balance a wheel on a bicycle so that it spins without wobbling or asked a mechanic to balance the wheels on a car, then you'll know that it doesn't take much unevenness in the distribution of mass to cause problems. A spinning structure, with humans wandering around inside it, will wobble and speed up, or slow down, every time anyone does anything, unless the mass of the structure is large enough to make those changes very minor.

All of these considerations led a number of researchers in the mid-twentieth century to propose a variety of very large space-based artificial habitats. Some of the most well-known versions are the so-called O'Neill cylinders, after Gerard O'Neill, a US physicist working in the 1960s and 1970s.[20] O'Neill famously presented a challenge to his students at Princeton, asking them whether the surface of a planet is really the right place for an expanding technological civilization. The students stepped up with a slew of ideas indicating that the answer was "no," and their efforts to understand the problem included proposals for giant rotating structures, which they and O'Neill would go on to develop further.

A full-blown O'Neill cylinder is a massive thing. In theory, it should be around 8 kilometers in diameter and 30 kilometers in length, a size where rotating around the long axis to mimic gravity at Earth's surface would require only 28 revolutions per hour. It is also not a single cylinder, but two, with one nested inside the other and spinning in the opposite direction of the outer one. This configuration cleverly enables a form of attitude control—essentially exploiting the rotational momentum and gyroscopic properties of the spinning structures. Attitude control is key because an O'Neill structure needs to point toward the Sun at all times in its heliocentric orbit so that solar power flows right down the cylinder through one open, or transparent, end. By forcing small changes between the long axes of the two spinning cylinders, the whole structure can be precisely maneuvered and pointed.

At this size, the 700 or so square kilometers of the interior of the inner cylinder could be turned into a habitat complete with its own atmosphere—this would also help shield against cosmic radiation. Critically, that atmosphere wouldn't be a uniform, pressurized gas (which could result in unmanageable pressure forces on the walls) but would settle toward the inner surface, like the way that a fluid runs up the sides of a spinning bucket. If the inner surface also held artificial lakes or even seas, there would be hydrological exchange with the atmosphere, and the interior volume could even see its own weather systems forming.

Not surprisingly, as serious as O'Neill was about the idea, it's also been great fodder for science fiction and speculation. Others have riffed on the basic design, with concepts like the Stanford Torus, which comes in the form of a skinny, doughnut-shaped structure more akin to a bicycle wheel.[21] The Russian cosmist Konstantin Tsiolkovsky had, of course, preempted most of these ideas back in the early 1900s, discussing the notion of rotating an environment to generate artificial gravity. He called this a "bublik city" after the torus-like shape of the Eastern European bublik bread roll, similar to a bagel. It's also not necessary for a spinning structure to be one continuous loop. A "bolas" with two massive, habitable pieces connected via a cable or a single inflexible spoke could also work and would save a lot of material.

The size of these habitats is a major consideration for more than just artificial gravity. Inside the atmospheric sheath of a rocky planet, there can be interleaved cycles of interacting gases and liquids that establish a dynamical stability, like a juggler balancing a spinning plate on a stick. Earth's climate is one such balancing act, and life here has evolved to be integrated with these cycles, which absorb and dampen a certain amount of change. New habitats that are independent of a planet and are large enough might be engineered to provide some of that same environmental self-regulation and be styled to match life's needs. That's more than can be said for a planet like Mars, which has its own firmly entrenched natural environment and a gravity well and open atmosphere that are, to be blunt, a pain to deal with.

Of course, being able to actually forge structures like these is another matter altogether. Even a smaller O'Neill cylinder would require several billion tons of materials, depending on the details of its construction and the tensile requirements. That's similar to the mass of a modest asteroid of about a kilometer in size, or a few times more massive than the estimated mass of all of New York City's buildings across its five boroughs. Consequently, an idea that has been debated for decades is to forgo some of the processing and refinement of materials to build a structure, and to instead drill into asteroids, hollowing out space as needed. If an asteroid is the right type—not a rubble pile but mostly solid rock—it would in principle be possible to spin it up along a preferred axis to generate the needed artificial gravity.[22]

However, this is a tricky business. If you make a hollowed cavity too big or spin it too fast, the material strength of the asteroid won't be enough, and it will fall apart. But calculations indicate that rotating asteroid habitats that are on the scale of at least a few hundred meters could indeed be made to work.

This is an appealing strategy for all kinds of reasons. Starter habitats could be attached on the exterior, or immediate interior, of a spun-up solid asteroid and need not be huge. As long as you're OK with walking around on what is, technically, the ceiling, your gravitational needs could be met on the spun-up body. Excavating more and more internal space could be an ongoing process and need not be rushed as other infrastructure is gradually built out. In essence, life on a habitat like this would grow and develop a bit like the way that mound-building termites engineer underground spaces, where they use the excavated materials to fabricate their large, temperature-controlled, and ventilated above-ground structures.

Engineering asteroids in space calls for new technological approaches to move or modify materials, and the more that the necessary resources and energy can be found locally, the better. For example, on the Earth, solar furnaces deploy arrays of mirrors to concentrate sunlight and achieve extreme temperatures as high as 3,500° Celsius (more than 6,000° F) to drive electricity generation or to melt and process metals or other compounds. Using sunlight this way in space is an intriguing possibility. Asteroid material like carbonate, or silicate minerals, will melt somewhere between 1,200 and 1,800° Celsius. A maneuverable solar concentrator could, in principle, easily melt or just sinter the dusty, crumbly material of many rubble-pile asteroids. Sintering avoids fully melting material but heats it to a point where it can be compacted and molded. By tracking the focal point of concentrated solar radiation across an asteroid, it could be either "sealed" or selectively heated and reformed into solid structural elements. This is an approach similar to the type of 3D printing known as stereolithography, which sets, or cures, photoreactive material out of a vat of liquid. It also resembles the technique of selective laser sintering, which forms three-dimensional structures from powdered materials.

If, instead, you wanted to fully disassemble or distribute the contents of a rubble-pile asteroid, it might be possible to attach ion-propulsion units and spin it up until centrifugal forces exceed the weak gravitational forces holding the body together. Another option might be to release a large cloud of gas on an orbital trajectory that crosses the path of a loosely held together asteroid and let the physics of gas drag disassemble and separate out the asteroid's component parts. From studying the mechanisms of planet formation, we know that solid particles that encounter interplanetary gas experience drag forces. These forces are acutely dependent on the size of the solid particles and will therefore sort the solids out by changing their orbital speeds according to how large they are.

All of these ideas can seem quite fantastical, and they gleefully ignore most of the details that might create temporary impediments to carrying out this kind of cosmic refurbishment. But the real point is that this is a direction that the future *might* take us in. All of these possibilities are just that, a type of "existence proof." They are evidence that there are no show-stopping laws of nature to prevent persistent living systems from resculpting cosmic terrain to pursue the same kinds of resources and living conditions that they have on the Earth, just farther afield.

In the nearer term, asteroid mining and repurposing could conceivably alleviate some of the burdens of pollution and energy use on the Earth that come from extracting precious elements to support our technosphere and human civilization. But the steps on the economic ladder to make that at all feasible are indeed significant. From the size of the launch masses of the first pieces of infrastructure from the Earth to the Delta-v's and the technical complexities of resource collection and refinement, the cost would be gigantic. It's not hard to imagine that such an effort, at least at the start, could contribute to even greater inequality between nations and people. On the other hand, if these efforts were ultimately in support of an entirely new, off-world economy, maybe that would produce different and more positive results, as we'll explore in the final chapter.

We might also continue to question whether biology has any real place beyond the Earth and whether we should look instead to robotics as a way to explore and provide access to the matter of the solar system. As with Mars, I think this is a valid philosophical stance, but I also think that it ignores life's relentless, mindless impetus to extend itself, and it ignores the exponential benefits of a synergistic ecosystem of both biological life and machines.

Sculpting the matter of an entire planetary system is a pretty grandiose thing, but it may be what life does, if it can. One main reason why this enterprise can seem like a spectacularly absurd pipe dream to us is because humans, as they are today, are probably not the end recipients. The scope of a future that would see us convert a billion asteroids into thousands of technospheres and biospheres is way, way off our scale of familiarity and is not something that any members of our present species can expect to experience. However, we've faced similar cognitive adjustments before, and the first step is often the hardest.

The puzzlement that Charles Darwin felt when confronted with the array of species and environments of the Galapagos Islands derived from a still incomplete worldview. Science had yet to reveal that Earth's deep time wasn't measured in thousands of years, but in billions. The trillions of microscopic experiments being carried out by life at every moment still eluded easy analysis; consequently, it was supremely difficult to comprehend how the forms and conditions of living things, scattered across the varied rocks of 127 islands, had arisen.

Now scatter living things and their constructions and robotic tools across a new archipelago orbiting the Sun on scales where everyday notions of causality and simultaneity fall apart: a landscape where gravity is no longer a singular thing that affects all beings on Earth the same but instead becomes a set of complicated choices and problems. It should come as no surprise that we're scratching our heads over whether this represents a plausible future in terms of physics, biology, or humanity. There might be a future out there that is free from scarcity or a future where the historical problems of inequity and short-term thinking still plague us. Or those problems could actually be diluted away by sheer scale as humanity fragments into what are, in effect, new species that radiate out across the solar system and are driven by a wealth of energy and materials.

─11─

THE DISPERSAL

We are therefore, driven to the conclusion, that causes generally quite inappreciable by us, determine whether a given species shall be abundant or scanty in numbers.

—Charles Darwin,
The Voyage of the Beagle

To understand our strange future, we have to remember that we are a dying species, as are all species on the Earth. But this condition is not just because of the inevitable geophysical and astrophysical changes that will modify and even erase the Earth's present environments. Our species, like any other, hovers in only a temporary—largely illusory—equilibrium before evolution whisks things along in a churn of divergence, speciation, and extinction. All of this has happened before, and all of this will happen again. The early hominins, who provided an opportunity for *Homo sapiens* to emerge into a distinct form, had their own extensive history across hundreds of thousands of years before dissolving away. Some left their genetic echoes in us all, others perhaps not. For most life-forms there is a Faustian bargain of sorts made with evolution. That bargain dictates how long any innovative and complex forms can have their days in the Sun before succumbing to evolutionary pressures and being subsumed into or replaced by new glittering creatures. But sometimes an innovation comes along that further modifies the rules of the game itself.

On the face of it, space exploration is such an unequivocally ridiculous trait for a species to acquire that it's absurd. Life that has found profound success for four billion years by integrating with—and modifying—the systems of a single wet, rocky planet is now pushing its boundaries into what seems like unspeakably inhospitable terrain. That move is fraught with problems, whether from the constraints of physical laws or the incompatibility of terrestrial ways of living with the energetic needs and damaging forces of the local cosmos. To make any of it work, extraordinary demands are placed on life's extended augmentations, in the form of robotic systems and engineered materials, as well as the neural systems of human designers and scientists. Missions are assembled that contain millions of physical parts, elaborately synthesized from Earth's diverse mixture of available elements. Altogether, these parts can have thousands of functions and are built to extraordinary standards of redundancy and flexibility, which anticipate events that may or may not ever be encountered. This is almost antithetical to biological evolution's frequently thrifty, no-frills efficiency.

Even more absurdly, space exploration has its roots in the nonmaterial, in the symbolic abstractions that people like Isaac Newton and Pierre-Simon Laplace invented to describe the motions and properties of the world. The moment that they formulated mathematical expressions to calculate an orbital path, they took the first steps toward being interplanetary. When Émilie du Châtelet, Mary Somerville, and Emmy Noether crafted their insightful interpretations and far-reaching theorems, they sped up this evolution of the possibilities of evolution, turbocharging it with ideas and clues. Darwin, with his nose deep in Earth's pungent biosphere, also changed how we think about the world and our path from present to future. He made the present more significant and the future far more open-ended and interesting.

As our species has explored these ideas, we've simultaneously extended life's tendrils across the solar system and turned the Earth into an enormously scrutinized and digitized world. In doing so, we've learned that there are no other planetary environments that match what exists on the Earth. The Moon and Mars are fascinating and important way stations, but they are also extraordinarily challenging places for life like us to exist, with their weak gravity and exposed environments. Venus and Mercury present brutal conditions for terrestrial life and materials, except perhaps for the racing Venusian cloud tops, whereas somewhere like distant Titan is so alien that it remains a place solely for machines. There may also be chemically rich, dark oceans inside icy moons and dwarf planets in the outer solar system, but for now they're barely accessible to us.

Yet life is busy making its transition to being interplanetary right now. It's very clear that for the past century we've been creating a substantial presence outside of the Earth. We are heading outward, learning as we go, and dragging the rest of Earth's living systems along with us. Through tens of thousands of rocket launches and clever, complicated machines, the evolutionary path of life on Earth has reached an inflection point and a new trajectory. One possible if not probable form of this trajectory is Dispersal, the true nature of which finally comes into view: a synthesis of space exploration's history, the architecture of the solar system, and the malleable nature of evolution itself, all projected into the future.

Once again, there are lessons from past experiences. When Darwin sailed on the *Beagle*, from 1831 to 1836, he had an opportunity to see a planet on the cusp of unimaginable change. In the 1830s the total human population was just one billion, the cumulative energy use by the technosphere was twenty-five times less than it is today,

and oil and natural gas hadn't yet become global resources or cata-
lysts of conflict. The estimated extinction rate of species in the 1830s
was ten times less than it is now. Global atmospheric CO_2 concentrations sat at around 284 parts per million, barely changed from a hundred years earlier, and far lower than today's 422 parts per million.
Atmospheric concentrations of the potent greenhouse gas methane
were around 780 parts per billion, compared to today's levels of over
1,850 parts per billion—a figure made even more significant because
methane has an atmospheric lifetime of just twelve years and persists
only because there are human actions that cause it to be constantly
released into the atmosphere.[1] Although there were plenty of wars
and geopolitical tensions in the 1830s, and across different continents
many awful practices of slavery and subjugation continued unabated,
truly global conflicts and weapons technology capable of fully rewriting the history of life on Earth were yet to come.

The magnitude of those subsequent changes is reflected in something that the cosmonaut Yuri Gagarin said more than a century later.
One of his reflections on being the first human to make a single complete orbit of the Earth, in April 1961, was that "looking at the earth
from afar you realize it is too small for conflict and just big enough
for co-operation."[2] This astute and heartfelt observation was made in
a radically different world than Darwin's. By the 1960s, Earth was
filled with interconnected humans pushing up against one another
and was a place where nations could now unleash the fury of thermonuclear bombs to wipe all civilizations off the planet.

But if we project a little way into the future, Dispersal could actually turn the human clock back in more positive ways. We know that
physical scales and temporal separations profoundly affect complex systems, as they have done for organisms scattered across places like the
Galapagos. If our species and other species become spread across the
solar system in a Dispersal, we will all begin to disconnect in ways that
are far more profound than what happens across Earth's great oceans

and continents. Dilute us enough, and there are parallels not just with the conditions of the 1830s but also with a time a hundred thousand years or more ago, when generation upon generation of hominins could exist without ever feeling the direct influence of other groups.[3]

Of course, it can't be quite the same, not least because of the webs of communications and transport that have to exist to keep interplanetary existence running. But this very same technological connectivity also creates an opportunity—if taken—to break away from the cycles of economics and inequalities that are so entrenched on the Earth. Further down this timeline there could be new forms of independence for humans and eventually new openings for biological evolution to take our species on thousands of new paths, especially if we tinker with biology directly, through gene editing to tweak our physiological baselines, driven by a desire to be more resilient to alien environments. All of this could make our descendants' future and the future of all life in the solar system far more robust against any kind of existential crisis. Dilution and diversity may be the ultimate strengths in a Darwinian universe.

If this all sounds too good to be true, well, of course, it might be. The path to all futures will inevitably have glitches and terrible traps. But Dispersal is a possibility, another kind of boundary condition in a complicated landscape of decisions and turning points. As with space exploration today, those possibilities will depend on ideas and on how science and technology evolve alongside us, and evolve they will.

There is no way for life on Earth to disperse into the cosmos without robotic augmentation and without that and other technologies being a constant, symbiotic presence. This is not surprising at all. Here on the Earth, many of us are entirely dependent on technology that is so omnipresent that it fades from view. How many of us really see the technosphere around us anymore? Even a farmer, on a remote plot of land on one of Earth's continents, is likely to work with tools forged by technologically derived metals and plastics, and use paths and roads that connect to a global web of transport. But they will seldom pause to question the sphere of the world that generated all of these things.

Dispersal forces the nature of these technological infrastructures to selectively change, becoming even more integrated with daily existence. It's a trajectory that we're already on. You and I, if we're lucky enough to live in certain privileged conditions, already have at our beck and call a constellation of supportive technologies. The windows and doors on our dwellings allow us light, movement, and climate control. We can communicate with people for help or for resources. We can get our food, medicine, clothing, and entertainment delivered. Electrical energy is conducted into our residences, drinking water appears out of pipes, and garbage and bodily waste are spirited away.

As individuals, many of us can trigger a flurry of activity in responsive machines and devices. From the moment that you hit Send on a screen, instructions cascade through a complex tree of algorithms and physical infrastructure. That bag of frozen peas you just ordered requires plants to have been grown, nurtured, and harvested, then packaged, shipped, delivered, and confirmed. Your one request influences the statistics of supply and demand for pea plants and farming, inducing future changes, no matter how subtle, into the economics and efficiencies of pea production. You're reaching across the planet to gather your food and influence the nature of food. Equally, to create your own monetary resources, whether you stock shelves, hammer nails, or copyedit text from home, there are few of your actions that don't pull the strings of a vast and unseen network of machines and avatars. That network is increasingly integrated with artificial intelligence and other forms of machine learning. The types of neural-net–mimicking AI we have today are extremely good at modeling vast quantities of data, whether that's human text and media or sensor and measurement information. These AI systems create their own extremely complex representations of these data and can then parse queries or predict outcomes rapidly and—usually—accurately. They can even generate instructions or plans designed to accomplish requested goals, whether in the form of navigational directions or business models.

In a Dispersal scenario, robotic, AI-governed devices would be key for handling the autonomy demanded by the isolation of light-travel time. In a medium-term forecast, these devices would exist in vast numbers—billions or trillions, in fact. Yet that would still only be a small percentage of the eighty thousand technospheres that could eventually be forged from the solar system's materials. Across interplanetary space, resources in matter and energy would occupy a fluid and adaptable web of orbits and relays. Imagine this example: A settler on Mars is informed by their habitat's AI that they will eventually need a new battery for one of their oxygen-handling systems.[4] So they (or the AI working on their behalf) will order it and set in motion a chain of events that connects a mining drone on a metal-rich asteroid in the asteroid belt to a constellation of orbital cyclers. These cyclers comprise a huge heliocentric cloud of general-purpose containers, refiners, and relays spread out in orbital ellipticity and distance—a constellation of interplanetary shipping lanes. Materials are extracted and exchanged, handed off to one cycler after another, until they reach a fabrication plant on (for the sake of this example) the Martian moon Phobos. The finishing manufacturing steps use a 3D-printing process to build a battery that is then slid into an atmospheric delivery drone that spirals down to Mars in a reusable hypersonic reentry carrier before disgorging itself to buzz the final few kilometers to make the delivery to our Martian settler. The time between ordering and receiving might be an Earth year, but the battery arrives precisely when it was expected to and when it is needed.

By today's standards, this simple delivery takes an extraordinary and impractical amount of resources. But in this future, it is routine and profitable. That's because the solar system's economy is now mea-sured in quintillions of dollars every Earth year, more than ten thou-sand times today's global economy of around $100 trillion per year. Ironically, the systems of on-demand manufacture and just-in-time delivery that have emerged in Earth's current global economy—and that are shockingly fragile in many ways—are precisely the style of systems that could underpin a far more robust Dispersal exo-economy.

Critically, even the very fundamentals of space exploration still have room to evolve in order to accommodate this kind of future. Take, for example, the story of what took place in 1990, when Japan launched its first lunar probe. The Hiten mission (named after mytho-logical beings who fly on clouds around the Buddha) was intended to reach a highly elliptical orbit around the Earth in order for the probe to swing close to the Moon, similar to a conventional Hohmann-style orbital transfer.[5] But something went wrong, and Hiten's orbit around the Earth was left woefully short of reaching the Moon, while its rocket had too little propellant left to fix things.

Two US scientists, Ed Belbruno and James Miller, then at the Jet Propulsion Laboratory, realized that their work might provide a rescue plan.[6] Belbruno had been developing mathematical and com-putational tools to explore what he called "weak stability boundary theory," the "fuzzy boundary" where orbits within systems of multi-ple massive bodies can take on open, wandering forms at the bound-aries between order and chaos. This is related to dynamics like the stable and unstable Lagrange points. In the case of the Hiten probe, calculations suggested that the spacecraft's small maneuvering thrust-ers could be used to very slowly climb away from the Earth. Instead of heading straight to the Moon, though, Hiten would find itself drift-ing along a wonky-looking path to a distance four times farther out, toward the Earth-Sun L2 point. The probe could then, in effect, fall back in a spiral to the Moon, where it would be naturally captured into lunar orbit in what's called a "ballistic capture," using no further fuel. The only drawback was the amount of time the maneuver would take.

This was like handing steering over to a sailor who reads the balance of currents and winds with such finesse that they can move a ship across an ocean without expending great effort, applying only some patience. In this case, it was about reading the complicated forces acting on Hiten because of the Earth, Sun, and Moon, all while these bodies were in their own orbital motion. When Belbruno and Miller were able to convince the Japanese space agency to try their proposal, it took the probe about three months to follow a complicated but extremely low-energy path that did indeed take Hiten into a successful lunar orbit and complete one of the most thrilling but underappreciated accomplishments of twentieth-century space exploration. By comparison, it took the Apollo astronauts around three days to make the same transition.

Beyond the Earth and the Moon, the solar system is a constantly shifting landscape of gravitational forces. With enough computing power, it is possible to expand these weak stability boundary calculations to map out what's been termed an Interplanetary Transport Network.[7] This is not a network of machines per se, but of these low-energy pathways that can, often through complicated and time-consuming trajectories, take spacecraft from one point to another with little expenditure of fuel and propulsion—by "reading the currents" of gravity. In 2001 a mission called Genesis was launched to capture particles from the solar wind and return them to the Earth. To accomplish this, Genesis tapped into a part of the Interplanetary Transport Network, threading a three-month-long path from launch out to the Earth-Sun L1 point (between us and the Sun), where it stayed in an enormous halo orbit, spanning nearly 3 million kilometers, for thirty months. The probe then took a convoluted, low-energy return path that saw it looping back all the way in the opposite direction toward the L2 point before coming back to reenter Earth's atmosphere in 2004.

The Interplanetary Transport Network works best in the vicinity of large planets and their moons because the transfer pathways tend to be less time-consuming. On the face of it, this doesn't sound great for humans getting around in the solar system; time is one thing that biology lacks. But for moving inanimate stuff, especially when a steady stream of materials or other items is in demand, the Interplanetary Transport Network might be an ideal solution. The shifting pathways of the network could carry trains of endlessly moving supplies to be tapped into as needed and refilled according to those needs, like the conveyor belts favored in some urban sushi restaurants. For instance, although an important plumbing part might take forty years to get to your base's broken toilet on Ganymede, you don't care because, for you, it was right there when you needed it, plucked out of the many items in an orbital carousel.

There are other aspects of space travel that have room to evolve, too. Lifting off a planet's surface into orbit is arguably the most challenging aspect of seeing a Dispersal get started at all. Here, too, there are options—ways to launch without launching. The most famous alternative to rocketry is an idea that, unsurprisingly, had its modern conception in the mind of the reliably creative Konstantin Tsiolkovsky in the late 1800s, and it consists of creating fixed structures that lead from a world's surface directly into space.

Tsiolkovsky imagined a tower tall enough to climb above the atmosphere, from where you could launch into an orbit with far less effort. The catch with this idea is not the principle but in the required strength and compressibility of materials. No presently known construction technique could make a tower hold up against its own weight if stacked as high as the outermost wisps of Earth's atmosphere, some

100 kilometers up. But in the 1950s and 1960s, other scientists caught on to this approach and came up with some brilliant variants.

The Russian engineer Yuri Artsutanov devised the concept of a geostationary "anchor" orbiting 36,000 kilometers above the equator. If a structure, like a cable, was lowered from the anchor, it could be attached to a fixed point on the Earth, and the tensile forces in the cable would both support its weight and cause its entire length to move with the same twenty-four-hour period instead of at the much higher orbital speed at each intervening altitude. Once you had this cable in place, you could literally climb up to space and down to the Earth, and at the geostationary anchor point you would already be in a perfectly stable orbit.[8]

It sounds so fantastically simple, but there are some catches. The first is the tension that parts of the cable would be under. Because only the geostationary anchor of the cable is strictly orbiting (and an orbit, you'll recall, is a condition of weightless free fall), the lower parts of the cable, moving at suborbital speeds, do feel their own weight, and for a uniform cable the weight and tension are at a maximum right before the geostationary anchor. One way to mitigate this is to extend the whole cable to even higher orbits, with a counterweight at the outer end. That counterweight will be pulled around at a faster-than-orbital speed for its height, and it will provide an additional force to help support the whole structure. This arrangement also shifts the center of mass of the system to be above the geostationary anchor point, reducing the bulk and strength requirements of the cable at that location.

In all scenarios, though, the cable material has to be as light as possible and with as high a tensile strength as possible, and it probably needs to vary in thickness and shape to adapt to the height-dependent tensions and conditions. But the necessary tensile strength is some 30 to 150 times greater than any materials that we can currently easily produce. Extremely strong cables or wires made with carbon nano-tubes might theoretically work because nanotubes are, in effect, giant molecules whose intramolecular bonds are extremely hard to break. But making lengths of tens of thousands of kilometers of nanotubes is not yet feasible. It's also not enough to just pin a cable between two points and stop worrying: that cable must cope with its passage through Earth's atmospheric zones. These zones include corrosive oxygen ions in near-Earth space and winds and weather lower down. There are also micrometeoroids that constantly zip through the near-Earth environment, gradually eroding material surfaces. Another hazard is of our own making: the constellations of satellites we've placed into orbits beneath a geostationary anchor. Even the varying gravitational tides of the Moon and Sun will add their forces to everything else that the cable is trying to withstand.

Those may be significant challenges, but they all seem to have solutions in better technology. That just leaves some engineering questions about exactly how to ride a cable up and down. The most attractive proposition is a self-propelled space elevator. But this also presents a number of new puzzles to solve, for any elevator car must climb and descend anywhere from hundreds to tens of thousands of kilometers. A trip from Earth's surface to the geostationary point in an elevator that zooms along at, say, 100 kilometers per hour might take two weeks to complete. The total change in gravitational potential energy between Earth's surface and a geostationary point is about 50 million joules per kilogram of mass. Consequently, to hoist a 70-kilogram human over two weeks would take a steady power input of around 2.9 kilowatts (equivalent to two or three microwave ovens running constantly). That doesn't seem so bad, except that the elevator car will have to provide life support and food for that one human as well as carry along the mechanisms that propel it up and down the cable, all of which is likely to make the mass of the elevator and its sole human occupant much larger. A metric ton (1,000 kilograms) would require an average power of around 41 kilowatts to make its two-week-long journey—assuming that every bit of energy goes into the climb and isn't lost in friction or other inefficiencies. That's a power use similar to that needed to accelerate a total of five heavy trains away from a station.

Space elevators for the Earth might need to be a long-term ambition. For the Moon or even Mars, the material and energy requirements are far less, though. For the Moon, calculations indicate that a cable "hanging" toward the Earth and reaching just past the Earth-Moon Lagrange L1 point between these worlds could be constructed with known technology like Kevlar fibers. The challenge here would be the length, at more than 50,000 kilometers. But the upside is that an elevator system would need much less power to drive up and down in the weaker lunar gravitational field.[9]

There are even more radical possibilities for leaving planetary surfaces. One of these is the rotating skyhook.[10] In this scenario, a massive orbital station in low-Earth orbit extends a relatively short cable a few hundred kilometers long that it spins around itself like a well-planted Olympic hammer thrower. On each rotation, the end of this cable dips down into the atmosphere, either just into the upper reaches or conceivably all the way down to the planetary surface. Payloads can then be hooked as the cable tip zooms past and will be carried or flung up into space, where they can be released into an orbit. Meanwhile, the anchor-point station regularly resets its own orbit after shedding momentum to payloads, readying for another round.

If that sounds a bit hair-raising, it is. A study by Boeing in 2001 estimated that a configuration capable of hooking payloads from a 100-kilometer-altitude drop-off point would involve the hook end of the cable traveling at more than 3 kilometers per second, or Mach 10, so the suborbital drop-off system would have to be a hypersonic transport capable of matching that pace.[11] Yet in terms of construction, skyhooks demand nothing more than some of the stronger materials we already know how to manufacture. Once again, this kind of mechanism could also be far easier to use on the Moon or Mars.

For propulsion systems, the ultimate goal would be to do away with the rocket equation and the depressing burden of having to accelerate your propellant mass as well as your spacecraft mass. For spacecraft and payloads that can withstand extremely high accelerations, there is the option of firing them like a bullet from a gun or propelling them using electromagnetic forces—a bit like the idea of magnetically levitated trains or rail guns, where a linear motor accelerates a payload using a ripple-like magnetic pulse along a track. The challenge is to reach high enough velocities to get to space without accelerating so hard that your launch vehicle and its contents are crushed in the process. That typically means limiting what kinds of things you can launch or building a very long linear motor to allow enough time to build up velocity. An average human can withstand up to four or five times the acceleration of gravity on Earth's surface (4 or 5 g), similar to the extremes of an amusement-park roller-coaster ride. But it's not comfortable for very long. A low-Earth orbital speed of around 7.8 kilometers per second could be reached in about 2.6 minutes if you accelerated at 5 g that entire time. Starting from a standstill, this would also mean traveling a linear distance of around 600 kilometers.

This is clearly not easy, especially because Earth's atmosphere is a major impediment to anything moving at these hypersonic velocities and is prone to causing the kind of metal-melting temperatures that probes experience when reentering the Earth's atmosphere from space. Once again, this is a system that might work much better on the Moon and perhaps on Mars, or on any smaller, airless bodies across the solar system. To go from the lunar surface to a stable orbit around the Moon by using a linear-motor launch system would involve about half a minute of 5 g acceleration along a track only 22 kilometers long. If you consider that we operate around 1.3 million kilometers of conventional rail track on the Earth today, 22 kilometers of specialized high-tech track on the Moon is perhaps not too difficult to imagine. To simply provide a conventionally propelled spacecraft an energy-saving boost toward space would take even less. Out in the asteroid belt, linear-motor magnetic launchers could be an easy way to supplement ion-propulsion systems by spitting out craft with a large initial Delta-v for any destination across the solar system.

For zero-propellant propulsion in open space, there is the option to use the pressure exerted by light as it interacts with a material surface. The utility of sunlight for helping missions like Mariner 10 or MESSENGER maneuver and orient themselves demonstrates the possibility of bigger things. Purpose-built solar sails have already been tested in space, where large reflective sheets can accumulate forces from solar photons to produce a steady acceleration to an attached spacecraft. The first interplanetary solar-sail demonstration was carried out by the Japanese Space Agency in 2010. A 200-square-meter sail of polyimide material was able to accelerate a probe called IKAROS and produce a Delta-v of a hundred meters per second over several months. The sail even had liquid-crystal panels built into it that could modify the reflective surface in real time, helping control the orientation of the spacecraft.[12] In 2024 a nine-meter-by-nine-meter technology-demonstration mission was launched on behalf of NASA from New Zealand by the commercial launch company Rocket Lab. This mission was to test out new foldable composite materials that form the "booms" holding the sail out. The whole sail unfurls from storage in a container that is about the size of a microwave oven.

One catch with solar sails is the size required to get substantial propulsion from solar radiation and the fact that the larger the sail is, the more mass it brings along for the ride. To accelerate a one-kilogram sail operating at Earth's distance from the Sun at a rate of a meter per second (about the same as a fit cyclist pedaling their bike hard) requires 100,000 square meters of highly reflective sail, which is equivalent to a square about 330 meters on a side. Fabricating and controlling a structure this large and keeping its mass small is, at present, an aspirational goal, but sailing with sunlight is a beautiful thing. It allows a payload to modify its orbit very precisely, to either drop inward in the solar system or head outward—not unlike an oceanic sailing vessel tacking into the wind or catching a trade breeze: all with free energy.

There are other options for enhancing the velocities achievable with light sails. Light concentrators in the form of mirrors or giant, flat optical lenses—using a technique known as Fresnel lensing—could be used to bombard sails with a higher density of solar photons than they'd otherwise receive. Or, with much greater energy cost, high-power laser arrays could avoid the need for sunlight altogether and might even be stationed on a planetary or asteroid surface, from where they can push sails waiting in space with short-lived but powerful pulses of light.[13]

Riding these fountains of light could be a useful way to launch sails and mass across interplanetary space. It's an approach that's also been studied for sending tiny interstellar probes, most notably by the Breakthrough Starshot project in its 2016 feasibility studies. Strategically positioned laser arrays in the solar system could also conceivably provide help in decelerating speeding craft, although this is a trickier proposition. There are even ideas that would use a powerful laser system like this to flash heat, or ablate, tiny pellets of material that would then launch off in a hypervelocity train moving in excess of 120 kilometers per second to hit an absorbent plate on a waiting spacecraft. That stream of momentum from the pellets would accelerate the vehicle and could push it to the outer planets in less than a year, instead of the decades that it currently takes.

Another piece of the puzzle of Dispersal lies in how structures and machines can be fabricated and configured in an exponentially growing and spreading environment. Building thousands of technospheres is certainly within the solar system's resources budget, as we've seen, but exactly how would any species do this? Some answers could lie in emerging technologies that are already dominating future thinking on manufacturing and construction. 3D printing—"additive layer manufacturing"—is a great example, where a combination of precision robotics, software, and materials science creates a new, generalizable approach to building objects layer by layer.[14] We already fly around the world (perhaps unwittingly) on planes equipped with 3D-printed metal-alloy jet-engine turbine blades. The company Rocket Lab 3D prints almost all parts of the engines for its Electron rocket launch system. Another company, called Relativity Space, claims to print 86 percent of the structure of its rockets. And astronauts onboard the International Space Station have experimented with printing specialized hand tools on demand, bypassing the need for expensive and time-consuming transport in the next cargo-supply run to the station.

Studies of 3D printing for space exploration reveal a whole range of intriguing possibilities. Raw feedstock for printing, which can be stored as granular pieces or powders, can do double duty as radiation shielding for astronauts during longer interplanetary trips. Many natural environments have the critical elements of hydrogen, carbon, oxygen, and nitrogen needed to synthesize polymers that are easily manipulated for printing almost any design. And if a 3D-printing system can churn out modular components, those components can in turn be assembled into almost any kind of structure, like Lego building blocks. For a bonus, these manufacturing techniques can enable the wholesale recycling of materials—grinding up broken or discarded parts and reconstituting them via printing them into something new.

Environmental information and communications are also central aspects of Dispersal. Remote satellite observations of the Earth are a testbed for the tools that we'd want to apply to other places. As we've seen, a world like Mars will need weather forecasts, land-use data, and an internet. The same is true for free-floating habitats or footholds on asteroids and icy moons, whose environments are buffeted by internal processes and by the solar weather that sweeps the system. Data pipelines are critical, and we've been getting better and better at making these: from the triumphs of transatlantic telegraphs all the way to encrypted, multiplexed bitstreams in microwaves and laser beams. Exquisite GPS data from satellites have transformed life on Earth, and descendant techniques will transform the way we disperse and explore the rest of the solar system, whether in new lunar and Martian positioning systems or in deep space, where the precise lighthouse rhythms of distant spinning neutron stars in the galaxy—the pulsars—can provide an immutable coordinate system for the interplanetary void. Because of the finite speed of light, the exact timing of these beats will shift to reflect a spacecraft's changing distance relative to these cosmic objects. Both China and the United States have used spaceborne instruments to test run navigation with pulsars by monitoring their millisecond beats in emissions of X-ray radiation, and can already demonstrate positional accuracy to within several kilometers.

I've already pointed out how there are also far fewer rules about where to place habitats or facilities when you're free from the confines of planetary surfaces. It just remains to find the zones of easy growth and evolution in the solar system. For example, orbiting in the plane of all the major planets may not be ideal. Instead, with some energetic cost, an orbit around the Sun could be greatly inclined, lifting habitats farther away from hazards like asteroids and interplanetary dust, and potentially farther away from some of the Sun's great coronal mass ejections that belch outward from around the solar equator.

Alternatively, living near larger natural objects—whether planets, moons, or asteroids—might still provide some advantages, such as a proximity to resources or to others of the species who chose to stick with surface existence. But orbiting a world directly, like today's space stations, is not the only option. There are trailing heliocentric orbits, where a habitat might follow or lead the Earth around the Sun, and there are Lagrange point locations, like those we already use for many robotic missions. We've seen how the Sun-Earth-Moon system offers a hierarchy of these delicate mooring points, some with their own orbital options, like the halo orbit that the James Webb Space Telescope loops itself around. But as an alternative, the Sun-Venus system also has Lagrange points as well as power advantages, with solar illumination that is almost exactly twice the intensity of that at the Earth.

A species might choose to avoid conventional closed orbits or equilibrium points altogether. Instead, they could nudge habitats or facilities onto some of those weak instability transfer pathways in the Interplanetary Transport Network to take an endless meandering tour of the solar system. In this century you might hang out around old Earth, but in the next you will thread through the outskirts of the Jovian system, all with minimal energy expenditure, using the gravitational hills and dips to carry you like a drifting raft.

Despite such radical departures from planetary life, a mindset that has colored ideas of space exploration in the past is one that sees Earth-like social and political orders mirrored across any dispersed populations. Science fiction has certainly contributed to this perception, with endlessly lazy—but narratively convenient—recasting of classical story arcs into operatic space settings. But Dispersal involves scales of distance and scales of resources unlike any that our species has encountered before. Interplanetary distances aren't just about the cost of transport; they change the nature of causality and communication, creating far more isolation and separation. Economic systems that are connected to a cosmic scale of resources in materials and energy pose entirely new questions about the outcomes for a dispersing species, questions that I don't think even the best modern economists can answer satisfactorily. Really, the one thing that we can make a halfway meaningful prediction about for a long-term Dispersal is that it will dilute and modify everything, from the hierarchies of governance and society to the fundamental selective pressures acting on biological and technological evolution. Dispersing is far more likely to generate vast diversity and change than it is to maintain a monoculture of organisms and ideas.

That is, to me, an extraordinarily hopeful thing. Visions of a more appealing future for life on Earth, where we live sustainably and in relative harmony, feel like a very, very hard reach these days. Unfortunately, there could be any number of dramatic resets here on the Earth: our industrial world may suffer a catastrophic collapse, or maybe one of those stubborn asteroids will stray across our orbital path. From such a reset, human life might emerge a little muted and more manageable, or perhaps we'd simply race to restore the status quo and standards of living we've become used to. By comparison, Dispersal could not only eventually offer new patterns for our lives and civilizations; it could also lead us to circumstances we might otherwise have never imagined.

The grand arc of Dispersal may seem utterly fictional, a fever dream that can surely never be realized. I can't tell you it will happen, but then neither can anyone guarantee you that it won't. The extraordinary thing is that we have been creating the foundations of this possible future for centuries, first with our mathematical interpretations of the functions of the physical world, in laws and principles from the likes of Newton and Laplace that the agile minds of people like du Châtelet and Somerville further codified, and then with the application of these concepts to a flood of rockets and spacecraft that have transformed our view of ourselves as well as our access to the farthest reaches of the solar system. Critically, the human participants in that story have never actually been that homogenous. People of all genders, people of different color and culture, have all played key roles.

There is something about the enterprises of space and exploration that transcends division and inequity if given a chance.

Living systems are also extremely good at taking opportunities where they arise. One of the most fundamental and rather surprising facts that we've learned about the Earth is just how pervasive and persistent life is. All evidence points toward a common lineage for everything alive today that reaches back across some four billion years in an unbroken chain. That's despite many major (and many more minor) episodes of disruption and extinction during which huge swaths of species were deftly clipped from the branches of related forms. That carnage has extended from the tiniest single-celled microbes to enormous beasts consisting of trillions upon trillions of complex cells united in mutually useful cooperation. Yet we are all still here, from one rootstock of living systems.

That persistence is in part because this one lineage has insinuated itself into—it seems—every known niche, as well as some niches that we're only just beginning to recognize. There are cells floating in Earth's atmosphere, in its oceans, and across the continents. Microbial life—the bacteria and archaea—swarm inside of and help make the planet's soils and landscapes. There are single-celled species that literally eat rock, using chemistry to drill pathways through even the freshest volcanic basalt. Organisms inhabit the oceans to their very bottom, with creatures in abyssal zones that live at pressures hundreds of times those on Earth's surface. Scattered across these depths are chemically rich, hot, hydrothermal vent systems where vibrant oases exist of microbes and codependent species of tubeworms, crabs, and many other organisms.

Since the 1970s, scientists have recognized that this is part of a hot, deep microbial biosphere within the Earth's rocky crust itself: a realm of "intraterrestrials." Today, there is emerging evidence of a parallel cool, wet, deep biosphere in watery deposits in fractured rocks and chambers many kilometers below continental surfaces. These places can be extraordinarily isolated from anything happening on the planetary surface, with water that's been unmixed for millions of years. Down here, though, the radioactive decay of isotopes of uranium and thorium can break apart water molecules in a process called radiolysis, generating molecular hydrogen, oxygen, and hydroxyl compounds. That molecular hydrogen is fuel for microbial species, enabling them to persist in these interior zones, forming alien ecosystems that we're only just learning about.[15]

All of these observations confirm that life is a ferociously tenacious phenomenon on a world like the Earth. And that, in many ways, is not at all surprising. Earth's life has presumably emerged because of the rich array of molecules, environments, and opportunities that you get with a rocky planet forming around a young star in a galaxy like ours. Whatever mix of Darwinian selection has taken place (and we assume that an awful lot has taken place) has inevitably honed the mechanisms of life so that it succeeds across precisely these kinds of environments. But that success is measured across many scales of time and space. It took over two billion years for complex multicellular life to really erupt on the Earth. That's a long wait time, yet we wouldn't call life on Earth a failed experiment. Similarly, if it takes four billion years for life to overcome the energetic barrier of gravity in order to begin to insinuate itself beyond one planet, that's entirely on point.

Ideas are the same. They tend to persist and evolve across time, and they sometimes emerge long after we might have naively expected them to. There are all kinds of reasons why. Sometimes, the moment and zeitgeist are just not right. In other cases, it's because it takes time for human minds to acclimatize to new circumstances and to the contrast between the past and the future.

By the end of 1835, the people on the *Beagle* were more than ready to get home, to retread their steps back to their launch point. But their pathway to a final berth was far from straightforward. From the Galapagos, the *Beagle* sailed across the Pacific to Tahiti and New Zealand before arriving in Australia in January 1836. Here, Darwin marveled at yet another distinctive branch of mammalian life in the continent's marsupial species. By April 1836, spirits were raised as the *Beagle* began to make its way back to the United Kingdom, closing the loop on its circumnavigation of the planet by stopping at Mauritius in the Indian Ocean and the Cape of Good Hope at the southern tip of Africa.

From there they sailed up to St. Helena (where Darwin was appalled at the disrepair of the house that Napoleon Bonaparte had died in) and the lonely Atlantic islands of Ascension, where—to Darwin's initial dismay—an extraordinary and unscheduled official diversion sent them back west, across the Atlantic once again, to the Brazilian state of Bahia to gather yet more survey data on longitudes. Finally, at long last, after several more eastward-bound island stopovers in the Atlantic, on October 2, 1836, the *Beagle* sailed into Falmouth dock on the southern Cornish coast of England.

In Darwin's diary entry for this day, he writes: "To my surprise and shame I confess the first sight of the shores of England inspired me with no warmer feelings, than if it had been a miserable Portugeese settlement." Five years at sea, exploring vast, alien places, and learning to live with the privations of a tightly packed ship would retrain anyone's brain and behavior. The snug harbor at Falmouth must have felt extraordinarily strange and small, especially with the knowledge that this was the end point, a collapsed singularity of future existence, after the global landscape of wonders that Darwin and the crew had spent years traveling through.

Darwin later reflected on the scope of the *Beagle*'s voyage in his published account: "The map of the world ceases to be a blank; it becomes a picture full of the most varied and animated figures. Each part assumes its true dimensions: continents are not looked at in the light of islands, or those islands considered as mere specks, which are, in truth, larger than many kingdoms of Europe. Africa, or North and South America, are well-sounding names, and easily pronounced; but it is not till having sailed for some weeks along small portions of their coasts that one is thoroughly convinced how large a portion of our immense world these names imply." It's small wonder that coming home felt a little underwhelming or that it would take the following decades for Darwin to adjust, gather, absorb, and refine his thoughts. The contrast between what awaited him in Falmouth harbor and the future of ideas revealed by his experiences could not be greater.

We now face our own contrasts among past, present, and future. Many of us have become quite unimpressed by or even oblivious to space launches and space exploration because we live in a world where so many other things clamor endlessly for our attention. Yet the extension of life and its devices into the cosmos has been on an accelerating trajectory for the past century. We may soon find our attention restored and ourselves surprised at just how readily life from Earth takes to its new opportunities. That doesn't mean that life's Dispersal in the solar system will happen quickly, but it seems likely to be just as persistent and pervasive as life's catalytic occupation of the Earth, and we appear to be the agents of this change.

Darwin was able to marvel at the seemingly endless abundance of life's forms and relationships as they unfurled on one small rocky planet. Evolution almost guarantees that humans, as we are now, are likely not the ultimate beneficiaries of cosmic Dispersal. Nonetheless, we might be the first to marvel at the new forms and relationships that life launches into as it becomes interplanetary.

ACKNOWLEDGMENTS

Writing a book about a topic as complex as space exploration means relying heavily on the products of many other writers, thinkers, scientists, engineers, explorers, and artists. It also involves making many decisions on what to include and what not to include. Space is (as a certain Captain James Tiberius Kirk once said) the final frontier, and that's a pretty grand thing, so there are very many events and facts that haven't made it into this book simply because of narrative flow and length, and I acknowledge all the remarkable people and episodes that got left out.

Immense gratitude goes to my literary agent, Deirdre Mullane of Mullane Literary, who patiently listened to me insist that a new book about space exploration was interesting well before my ideas took proper shape. From lunch on a front porch in the Hudson Valley to many calls and emails, we managed to wrestle this beast into a compelling proposal. Thomas Kelleher of Basic Books not only gave us a home but also a bounty of critical insight and thoughtful feedback throughout the process, and Gillian Sutliff of Basic rounded the figurative decimal places with marvelously helpful and on-point editorial suggestions.

Scientific colleagues and intellectual confidants provided support and sounding boards throughout, in ways both known and unsuspecting, so thank you: Sean Sakamoto, Frits Paerels, Eric Gotthelf, Piet Hut, Mary Voytek, Mary Beth Wilhelm, Mike Wong, Olaf Witkowski, Ryan Felton, David Spiegel, Rebecca Oppenheimer, and Szabolcs "Szabi" Marka. Family and friends were a bedrock of solace as I wandered the depths of writer's space—my loving thanks to the inner circle of Bonnie, Laila, and Amelia. And thank you Suki, David, Steve, Carolyn, and Tim, along with friends and neighbors Abbie, Lewis, Mal, Ellen, Sean, Christine, Joe, Lorraine, TC, Carrie, Cigal, Uri, Kyra, Ilhan, Adam, Phil, and Claire.

NOTES

CHAPTER 1: OCEANS AND VESSELS

1. The text commonly referred to as *The Voyage of the Beagle* was published as C. R. Darwin, *Narrative of the Surveying Voyages of His Majesty's Ships Adventure and Beagle, Between the Years 1826 and 1836, Describing Their Examination of the Southern Shores of South America, and the Beagle's Circumnavigation of the Globe* (London: Henry Colburn, 1839). Here and throughout, I've made use of the digital version of Darwin's diary at Darwin Online (https://darwin-online.org.uk), which is reproduced from R. D. Keynes, ed., *Charles Darwin's Beagle Diary* (Cambridge: Cambridge University Press, 2001).

2. The next few billion years of the Earth's evolution are not fully understood, but fundamental changes in climate state and water loss, and the possible later engulfment of the planet, are expected because of the Sun's evolution. See, for example, Christine Bounama, Siegfried Franck, and Werner von Bloh, "The Fate of Earth's Ocean," *Hydrology and Earth System Sciences* 5, no. 4 (2001): 569–575; and K.-P. Schröder and Robert Connon Smith, "Distant Future of the Sun and Earth Revisited," *Monthly Notices of the Royal Astronomical Society* 386, no. 1 (May 1, 2008): 155–163.

3. The complete timeline of life on Earth is uncertain, but the oldest plausible evidence for living systems comes from around 3.5 billion years ago.

4. There is some dispute over the total because more nations have developed launch systems, but today eleven nations appear to have active launch ability. See, for example, Katharina Buchholz, "The Countries Capable of Launching Space Rockets," *Statista*, July 18, 2022, www.statista.com/chart/27792/countries-capable-of-launching-space-rockets. See also Alexandra Witze, "2022 Was a Record Year for Space Launches," *Nature*, January 11, 2023, www.nature.com/articles/d41586-023-00048-7.

5. Forty-eight countries and counting: "List of Space Travelers by Nationality," *Wikipedia*, https://en.wikipedia.org/wiki/List_of_space_travelers_by_nationality.

6. See, for example, Louis de Gouyon Matignon, "The Nigerian Space Program," *Space Legal Issues*, November 24, 2019, https://web.archive .org/web/20210510193501/https://www.spacelegalissues.com/the-nigerian -space-program.

7. See Indian Space Research Organisation (ISRO), www.isro.gov.in.

8. Alex Travelli, "The Surprising Striver in the World's Space Business," *New York Times*, July 4, 2023, www.nytimes.com/2023/07/04/business/india -space-startups.html.

9. The machinery of war and space exploration have a close association, but it doesn't account for all the efforts and technological developments. A good overview is contained in Neil deGrasse Tyson and Avis Lang, *Accessory to War: The Unspoken Alliance Between Astrophysics and the Military* (New York: W. W. Norton, 2018).

10. Satellites come and go, so the numbers are significantly variable, especially because large complements toward new constellations can come from single launches. Some numbers can be found at "Space Environment Statistics," European Space Agency, https://sdup.esoc.esa.int/discosweb/statistics.

11. Limiting the potential of modern astronomical observatories is not a joke at all. See, for example, Olivier R. Hainaut, "Large Satellite Constellations and Their Impact on Astronomy," European Southern Observatory, www .eso.org/~ohainaut/satellites; and Olivier R. Hainaut and Andrew P. Williams, "Impact of Satellite Constellations on Astronomical Observations with ESO Telescopes in the Visible and Infrared Domains," *Astronomy and Astrophysics* 636 (2020), https://doi.org/10.1051/0004-6361/202037501.

12. The archetypal photograph of the whole Earth was taken from the Apollo 17 mission in 1972 en route to the Moon, during the translunar-coast phase, by the crew: astronaut Eugene A. Cernan, commander; astronaut Ronald E. Evans, command module pilot; and scientist-astronaut Harrison H. Schmitt, lunar module pilot. The image was centered on the eastern coast of Africa.

13. Earth's atmosphere does not have a sharp "edge"; it becomes progressively less dense with altitude and continues thousands of kilometers into "space" as the "exosphere" that fades into the (still not empty) interplanetary vacuum. The 100-kilometer figure is a convenient benchmark and is known as the Kármán line, after the Hungarian-American physicist and aerodynamic pioneer Theodore von Kármán. This line is considered the "official" divider between Earth and space for spaceflight.

14. See, for example, Charles S. Cockell, ed., *The Institutions of Extraterrestrial Liberty* (Oxford: Oxford University Press, 2023). From the publisher's

description of this collection: "The exploration of space raises new problems in the expression of human freedoms. While the potential to establish new extraterrestrial settlements is thrilling, it also brings along a myriad of decisions to consider when addressing how these settlements should operate in a way which maintains human liberties."

15. The *Beagle* had a rich history before its 1831 voyage, including some significant rebuilds because this class of ship had a reputation as a "floating coffin." These rebuilds were further augmented by Fitzroy's careful additions. See Keith Thomson, "H.M.S. Beagle, 1820–1870," *American Scientist* 102, no. 3 (May–June, 2014), www.americanscientist.org/article/h-m-s-beagle-1820-1870.

16. For a description of Mount Edgcumbe, see www.visitgardens.co.uk /gardens/mount-edgcumbe.

17. An extraordinary moment during Darwin's five-year voyage with the *Beagle* involved a fossil tooth and the deep history of life on Earth. His own words set up the story: "November 26th—1833 I set out on my return in a direct line for Monte Video. Having heard of some giant's bones at a neighbouring farm house on the Sarandis, a small stream entering the Rio Negro, I rode there accompanied by my host, and purchased for the value of 18 pence, the head of an animal equalling in size that of the hippopotamus. . . . The people at the farmhouse told me that the remains were exposed by a flood having washed down part of a bank of earth. When found, the head was quite perfect, but the boys knocked the teeth out with stones, and then set up the head as a mark to throw at." Darwin then had a quite remarkable stroke of informed luck: "By a most fortunate chance, I found a perfect tooth, which exactly fits one of the sockets of this skull, embedded by itself on the banks of the Rio Tercero, at the distance of about 180 miles from this place." See this account of the Toxodon (rodent) tooth: Raphaëlle Costa de Beauregard, "Travelling in Darwin's *Voyage of the Beagle* (1839)," *Caliban*, https://journals.openedition.org/caliban /3768.

18. Influencing the genetic inheritance involves a lot more than "by-eye" breeding of species and ranges from medicine and agriculture to conservation and preservation. See, for example, Andrew P. Hendry et al., "Evolutionary Principles and Their Practical Application," *Evolutionary Applications* 4, no. 2 (March 2011): 159–183.

19. There is a deeper story here with many still-unanswered questions. See, for example, my work on this in Caleb Scharf, *The Ascent of Information: Books, Bits, Genes, Machines, and Life's Unending Algorithm* (New York: Riverhead, 2021).

CHAPTER 2: A HISTORY OF PROPULSIVE IDEAS

1. Escape velocity has a very precise mathematical meaning. It is the outward velocity that is required (starting right at the surface of a massive planet, for example) so that by the time you have traveled an infinite distance from your start point, your relative velocity to that start point will have diminished to zero. In other words, it is the initial velocity that will let you coast away for an infinite distance, even as the gravitational forces between you and your launch planet decelerate you. In practical terms, though, in a universe containing many massive objects, and because gravitational forces decrease as the inverse square of distance, you'll effectively escape the dominant pull of your launch planet in much less than an infinite distance or infinite time.

2. The energy density of life makes for some surprising comparisons. For instance, the total energy *produced* per unit time in the core of the Sun is about 275 watts per cubic meter, which doesn't seem so high and is even comparable to what terrestrial organisms produce. But in the Sun's core, that power has a hard time going anywhere because of the extreme density of matter (twenty times denser than iron), so the total—*cumulative*—energy density present at any moment in any region is very high.

3. The power flowing through Earth's biosphere is a tricky quantity to evaluate, but there are robust estimates. See, for example, T. M. Hoehler, D. J. Mankel, P. R. Girguis, T. M. McCollom, N. Y. Kiang, and B. B. Jørgensen, "The Metabolic Rate of the Biosphere and Its Components," *Proceedings of the National Academy of Sciences* 120, no. 25 (2023): e2303764120. See also G. R. Quetin, L. D. L. Anderegg, A. G. Konings, and A. T. Trugman, "Quantifying the Global Power Needed for Sap Ascent in Plants," *Journal of Geophysical Research: Biogeosciences* 127 (2022): e2022JG006922.

4. Estimates vary, and they need to be seen in the context of all of Earth's systems. See, for example, M. Williams, J. Zalasiewicz, P. Haff, C. Schwägerl, A. D. Barnosky, and E. C. Ellis, "The Anthropocene Biosphere," *Anthropocene Review* 2, no. 3 (2015): 196–219, https://doi.org/10.1177/2053019615591020.

5. I make this caveat explicit because very few scientific stories involve only one individual, but it simplifies the telling to pick some points of focus.

6. Émilie du Châtelet was an extraordinary and fascinating person. There are many sources about her life and works, but an indispensable one is Robyn Arianrhod, *Seduced by Logic: Émilie Du Châtelet, Mary Somerville and the Newtonian Revolution* (New York: Oxford University Press, 2012).

7. The 1687 *Principia* was not Newton's only work, but it is "the one" that we often think of when talking about Newton. Its formal title was *Philosophiæ*

Naturalis Principia Mathematica (*The Mathematical Principles of Natural Philosophy*) and it consisted of three volumes covering the general motion of bodies and gravitation, the resistance of motions by materials and mediums, and the systems of the world, which include the motion of planets and moons. Centuries later, it had a somewhat unjustified reputation as being opaque, but the truth was more complex. See, for example, Caleb Scharf, "Who Said Nobody Read Isaac Newton?," *Nautilus*, January 20, 2021, https://nautil.us/who-said-nobody-read-isaac-newton-238105.

8. Later in the 1700s, Joseph-Louis Lagrange realized that the concepts of kinetic energy and potential energy could make the calculations of any system dynamics (like a planetary system or a mechanical system) that much easier. This process uses a mathematical construct called the "Lagrangian" that is often just the equation for the difference between kinetic energy and potential energy.

9. See, for example, Carolyn Iltis, "Leibniz and the Vis Viva Controversy," *Isis* 62 (1971): 21–35. See the following article for more about the development of concepts of energy, du Châtelet, and further references: "The History of the Concept of Energy and Work," IOP, https://spark.iop.org/history-concept-energy-and-work.

10. Willem Jacob 's Gravesande was a Dutch natural philosopher and mathematician who, in 1722, published results on his experiment with brass balls dropped into soft clay, which indicated that the size of the indentations in the clay were proportional to the mass of the balls and the square of the final velocity of the balls—i.e., proportional to what would become known as the kinetic energy. He communicated these results to du Châtelet. See also Dora Musielak, "The Marquise du Chatelet: A Controversial Woman of Science," *arXiv*, 2014, https://arxiv.org/abs/1406.7401.

11. See Kathryn A. Neeley, *Mary Somerville: Science, Illumination, and the Female Mind* (Cambridge: Cambridge University Press, 2001); Mary Somerville, *Personal Recollections, from Early Life to Old Age* (Boston: Roberts Brothers, 1874); and Mary Somerville, *Mechanism of the Heavens* (London: John Murray, 1831), https://archive.org/details/mechanismofheave00somerich/mode/2up.

12. All four volumes can be viewed as scans of an 1829–1839 edition at the Smithsonian's online digital library: https://library.si.edu/digital-library/book/mecanique-celeste.

13. To modern ears the Society for Diffusion of Useful Knowledge sounds like a quirky British sitcom, but it was founded through the egalitarian ideals of the Whig political party, which would become the UK's Liberal Party.

14. The use of the theory of perturbations, also known as Laplace-Lagrange secular theory, extends to the question of the orbital stability of the entire solar system, a topic nicely outlined by the dynamicist Jacques Laskar here: "Is the Solar System Stable?," *arXiv*, 2012, https://arxiv.org/pdf/1209.5996.

15. Despite some professional recognition during her lifetime, it is only more recently that Noether's major contributions and insights have captured popular attention. See, for example, Dwight E. Neuenschwander, *Emmy Noether's Wonderful Theorem*, rev. ed. (Baltimore: Johns Hopkins University Press, 2017).

16. Newton's *A Treatise of the System of the World* was originally a part of his *Principia* but was translated into English and, because it was rather more accessible to lay readers, was published separately in 1728 with edits to make it seem like a separate work: I. Newton, *A Treatise of the System of the World* (United Kingdom: F. Fayram, 1728).

17. So if you fall in a straight line, is that an orbit? It could be thought of as a specific class of orbit known as a radial orbit, an orbit with the highest possible ellipticity in which—if matter didn't block your way—you would just oscillate back and forth around the center of mass of the system (which is something that can happen to stars in gravitating clusters).

18. A consequence of the nature of gravity as an "inverse-square law" force and the conservation of energy. This formula later became known as the Virial Theorem after work done in the 1800s.

19. See, for example, the information and diagrams on Delta-v budgets in the solar system here: "Delta-v Budget," *Wikipedia*, https://en.m.wikipedia.org/wiki/Delta-v_budget.

20. Walter Hohmann published this concept in his 1925 book *Die Erreichbarkeit der Himmelskörper* (*The Attainability of Celestial Bodies*), translation at https://archive.org/details/nasa_techdoc_19980230631. The Hohmann transfer is one staple tool in the repertoire of spacecraft orbital design, working well for moving to higher orbits around an object (e.g., the Earth or the Sun) or to lower orbits. Different types of Hohmann orbits involve different distances traveled around the elliptical transfer orbit, from less than 180 degrees to more than 180 degrees.

21. Konstantin Tsiolkovsky was quite a remarkable character, and his technical insights caused his work to become a point of pride for the USSR. This NASA translation of a Soviet collection of some of Tsiolkovsky's works gives a taste of the breadth of his contributions: *Collected Works of K. E. Tsiolkovsky*, vol.

2: *Reactive Flying Machines*, 1954, https://epizodsspace.airbase.ru/bibl/inostr
-yazyki/tsiolkovskii/tsiolkovskii-nhedy-t2-1954.pdf.

22. See, for example, this discussion of Moore's 1813 publication: W. Johnson, "Contents and Commentary on William Moore's *A Treatise on the Motion of Rockets and an Essay on Naval Gunnery*," *International Journal of Impact Engineering* 16 (1995): 499.

23. We already know that this is possible and have sent spacecraft on their way. For example, Voyager 2, now traveling at 15 kilometers per second relative to the Sun, should pass within about 1.7 light-years of the star Ross 248 in about forty thousand years.

24. Many accounts exist; a quick—but quite thorough—guide is available online from Embry-Riddle Aeronautical University: "History of Rockets and Space Flight," https://eaglepubs.erau.edu/introductiontoaerospaceflightvehicles /chapter/history-of-space-flight.

25. In 1919 Goddard published what is widely considered as one of the first practical treatises on high-altitude rocketry and reaching space: "A Method of Reaching Extreme Altitudes," https://edan.si.edu/transcription/pdf_files/8542 .pdf.

26. In 1920 the *New York Times* published an anonymous editorial stating that Goddard was wrong and that rockets could not work in the vacuum of space because they required air to push against to achieve thrust (evidently, the *Times* wasn't big on physics). In 1969, long after Goddard's death in 1945, but the day after Apollo 11 launched for the Moon, the newspaper published a "correction" that ended with "Further investigation and experimentation have confirmed the findings of Isaac Newton in the 17th Century, and it is now definitely established that a rocket can function in a vacuum as well as in an atmosphere. The Times regrets the error."

27. Few stories in the annals of rocketry or space exploration can match that of Jack Parsons. See, for example, Alex Lin, "The Occult History Behind NASA's Jet Propulsion Laboratory," *Supercluster*, November 10, 2020, www .supercluster.com/editorial/the-occult-history-behind-nasas-jet-propulsion -laboratory.

28. Oberth's paper describing this effect was published in 1929 as "Ways to Spaceflight," available at https://archive.org/details/nasa_techdoc _19720008133. It would be remiss to not point out that Oberth went on to work for the Nazis and for Werner von Braun. A more elaborate mathematical treatment of the Oberth effect is available in P. Rodriguez Blanco and C. E.

Mungan, "Rocket Propulsion, Classical Relativity, and the Oberth Effect," *Physics Teacher* 57 (2019): 439, https://doi.org/10.1119/1.5126818.

CHAPTER 3: THE HIGH PLAINS

1. The amount written about this endeavor is seemingly without limit. Some facts and figures (including the estimate of 650 million people watching across the globe) are given here: "Apollo 11 Mission Overview," NASA, www.nasa.gov/history/apollo-11-mission-overview. The full Lunar Excursion Module (LEM, or the Eagle) documentation and the command-module documentation are available at "Apollo Lunar Module Documentation," NASA, www.nasa.gov/history/alsj/alsj-LMdocs.html. The *Apollo Lunar Surface Journal*, compiling transcripts and other information, is here: www.nasa.gov/history/alsj. Many people directly involved wrote excellent books about Apollo and their own experiences. A particularly good one is by Apollo 11 astronaut Mike Collins, *Carrying the Fire: An Astronaut's Journeys* (New York: Cooper Square, 1974).

2. A variety of information on the Apollo 11 lunar spacesuits is available: "Apollo/Skylab Spacesuit," *Wikipedia*, https://en.wikipedia.org/wiki/Apollo/Skylab_spacesuit. A particularly good source is this history and "brochure" from the International Latex Corporation: "Apollo Space Suit," www.asme.org/wwwasmeorg/media/resourcefiles/aboutasme/who%20we%20are/engineering%20history/landmarks/apollobr.pdf. The suit's life-support system is described in this forty-page review: Kenneth S. Thomas, "The Apollo Portable Life Support System," NASA, www.nasa.gov/wp-content/uploads/static/history/alsj/ALSJ-FlightPLSS.pdf.

3. The electric lunar rover was as much a technological marvel as the spacecraft that took its different versions to the Moon. If you have a rainy day to spare, there is extensive documentation to read on every detail of these vehicles: "Apollo Lunar Roving Vehicle Documentation," NASA, www.nasa.gov/history/alsj/alsj-LRVdocs.html.

4. Long Island geological history: Kathleen M. Fallon, "Long Island's Dynamic History," New York Sea Grant, September 7, 2021, https://storymaps.arcgis.com/stories/4bf3df46568f4f3e93d222ac6ae08fee.

5. For an interesting tidbit about Luna 3 and the first Soviet images of the lunar farside using captured US radiation-hardened film, see Alan Bellows, "Faxes from the Far Side," *Damn Interesting*, October 2015, www.damninteresting.com/faxes-from-the-far-side.

6. The Ranger 3 and 5 impact landers (Ranger 4 did impact, unintentionally, on the farside) were clever bits of engineering, with balsa an excellent choice of material. Some technical details are available at "Ranger 3," NASA, https://nssdc.gsfc.nasa.gov/nmc/spacecraft/display.action?id=1962-001A.

7. Alexei Leonov's story, as well as the story of the US/Soviet race to space, is recounted in David Scott and Alexei Leonov, *Two Sides of the Moon: Our Story of the Cold War Space Race* (New York: Thomas Dunne, 2004).

8. The Apollo 15 and 16 subsatellites were innovative experiments to learn more about the conditions of lunar space. See NASA, https://science.nasa.gov/mission/apollo-15-subsatellite, and https://science.nasa.gov/mission/apollo-16-subsatellite.

9. The lunar mascons are generally thought to be the result of massive asteroid impacts that occurred during the first few hundred million years of the formation of the solar system. In 2015 data were published from a pair of gravity-sensing lunar satellites (the GRAIL mission) that further strengthened this theory. See G. A. Neumann et al., "Lunar Impact Basins Revealed by Gravity Recovery and Interior Laboratory Measurements," *Science Advances* 1, no. 9 (2015): e1500852.

10. George Darwin published an epic mathematical tome on tides based on lectures he delivered in 1897 at the Lowell Institute in Boston, Massachusetts. See *The Tides and Kindred Phenomena in the Solar System* (London: J. Murray, 1898).

11. The formation of the Moon is still not fully understood, but several leading hypotheses focus on the idea of a cataclysmic collision between the proto-Earth and another planet-sized object. See, for example, A. N. Halliday, "The Origin of the Moon," *Science* 338 (2012): 1040–1041, https://doi.org/10.1126/science.1229954. The extreme similarity of isotopic abundances in lunar material and in Earth material is almost too much to be explained by even this giant impact unless the impactor was formed essentially from the same protoplanetary material as the Earth. See S. G. Nielsen, D. V. Bekaert, and M. Auro, "Isotopic Evidence for the Formation of the Moon in a Canonical Giant Impact," *Nature Communications* 12, no. 1817 (2021), https://doi.org/10.1038/s41467-021-22155-7.

12. In total, 324 distinct permanently shadowed regions (PSRs) are currently identified on the Moon, all within about 11 degrees latitude of both the south and north lunar (geographic) poles. They vary in area from 10 square kilometers to more than 1,000 square kilometers.

13. The evidence for water ice on the Moon has been accumulating for a few decades. The Clementine mission in 1994 suggested the possibility. The Lunar Prospector mission in 1998 found evidence of excess hydrogen that could be in water molecules in shadowed polar regions. This led to a reassessment of Apollo samples with modern analysis tools that found hydrogen inside volcanic glass beads, indicating the early presence of water in lunar volcanos. The LCROSS impactor and Lunar Reconnaissance Orbiter observations in 2009 confirmed that there were grains of nearly pure water ice lofted skyward during the deliberate LCROSS impact on the Cabeus crater. And in 2018, data from the Moon Mineralogy Mapper, carried by ISRO's Chandrayaan-1, provided the first high-resolution map of minerals that make up the lunar surface and showed multiple locations of water ice in permanently shadowed regions of the Moon. A review (up to 2010) of the situation on water on the Moon is in M. Anand, "Lunar Water: A Brief Review," *Earth Moon Planets* 107 (2010): 65–73, https://doi.org/10.1007/s11038-010-9377-9.

14. Helium-3 isotopes are often cited as one of the resources that countries may compete for on the Moon. But the very, very long timeline of yet-to-be-viable fusion-technology development suggests that whether or not substantial helium-3 is on the Moon, it won't matter until far in the future.

15. The Apollo Guidance Computer was a landmark development. If you want to try your hand at using it, there are online emulators that can give you a sense of what the astronauts had to learn. See https://svtsim.com/moonjs/agc.html or www.ibiblio.org/apollo/#gsc.tab=0 for these and other space-mission computer emulations.

16. Code development for Apollo was indeed a huge, groundbreaking task. If you're interested in seeing what this looked like, some partial Apollo command code software has been reconstructed from hard-copy printouts used at the time. See, for example, https://github.com/chrislgarry/Apollo-11. (Warning: for hardcore coders only!)

17. The transport of biogenic oxygen from the Earth's upper atmosphere to be implanted on the surface of the Moon is an extraordinary phenomenon and does seem to be supported by the data. See, for example, S. Perkins, "Earth Is Sending Oxygen to the Moon: Charged Particles Swept to Our Satellite May Explain Lunar Soil Oddities," *Science*, 2017, https://doi.org/10.1126/science.aal0674; and K. Terada et al., "Biogenic Oxygen from Earth Transported to the Moon by a Wind of Magnetospheric Ions," *Nature Astronomy* 1, no. 0026 (2017), https://doi.org/10.1038/s41550-016-0026.

CHAPTER 4: THE TERRANS

1. After Germany's defeat in World War II, the United States moved quickly to "recruit" (one can hardly imagine there was much reticence) more than six thousand German scientists and engineers, and to relocate technology such as the advanced V2 liquid-fueled rockets. A hundred of these rockets were shipped to the White Sands missile range in New Mexico, and the first nominal launch was in May 1946. In October 1946 a launch reached an altitude of 105 kilometers and returned film with photographs of the Earth—the first images ever taken from space.

2. The first fruit flies in space were actually launched in 1946 but were not recovered. The 1947 flight was the first successful recovery of (unharmed) fruit flies.

3. For more information about early animal experiments in space, see Colin Burgess and Chris Dubbs, *Animals in Space: From Research Rockets to the Space Shuttle* (Berlin: Springer, 2007), https://doi.org/10.1007/978-0-387-49678-8.

4. Despite Valeri Polyakov's long, arduous stay in space on Mir (plus another eight months in orbit on other missions), he lived to be eighty years old. See "Valeri Polyakov," *Wikipedia*, https://en.wikipedia.org/wiki/Valeri_Polyakov.

5. You can read *The Brick Moon, and Other Stories* by Edward Everett Hale at www.gutenberg.org/ebooks/1633. The story was originally published in serialized form in *The Atlantic* beginning in 1869.

6. Interesting documentation about the Salyut space-station program can be found here: "The Salyut Era: First Space Stations," www.russianspaceweb.com/spacecraft_manned_salyut.html.

7. On the impacts on human health of microgravity and other aspects of being in space, see C. Krittanawong et al., "Human Health During Space Travel: State-of-the-Art Review," *Cells* 12 (2023), https://doi.org/10.3390/cells12010040. An entertaining and highly informative book about the trials of human biology in space is Mary Roach, *Packing for Mars: The Curious Science of Life in the Void* (New York: W. W. Norton, 2010), https://doi.org/10.3357/ASEM.2972.2011.

8. M. Stavnichuk et al., "A Systematic Review and Meta-analysis of Bone Loss in Space Travelers," *Nature Microgravity* 6 (2020), https://doi.org/10.1038/s41526-020-0103-2.

9. Explorer 7 consisted of two cone-shaped halves connected at their widest ends by a cylindrical drum. The satellite's primary scientific goals were actually

to measure solar X-rays and ultraviolet light along with high-energy cosmic rays and micrometeoroid impacts as they punctured thin sensors on the spacecraft surface. The visible and thermal radiation of the Sun and Earth was very much a secondary experiment attached to the spacecraft bus. Although Explorer 7 died in 1961, it still orbits the Earth. A description of this experiment from 1959 can be found at "Explorer 7: (1959 IOTA 1)," https://digital.lib.uiowa.edu/islandora/object/ui%3Avanallen_2938/datastream/OBJ/view.

10. "Thermal Radiation," NASA, https://nssdc.gsfc.nasa.gov/nmc/experiment/display.action?id=1959-009A-01.

11. A description of the TIROS satellite series can be found here: "TIROS," NASA, https://science.nasa.gov/mission/tiros. Interestingly, the very first downloaded images from TIROS-1 had to be processed, printed, and flown first by helicopter to the nearby county airport from the ground tracking station in Fort Monmouth, New Jersey, and then by plane to Washington, DC, to be delivered to NASA headquarters, although poorer-quality versions were sent via an early fax system. See also "Milestones," https://ethw.org/Milestones:TIROS-1_Television_Infrared_Observation_Satellite,_1960.

12. Higher-altitude clouds of water vapor in Earth's atmosphere are cold and are typically shown as white in infrared images, whereas lower-altitude clouds tend to be warmer and show up in gray tones. Consequently, infrared images can help identify cloud heights as well as coverage.

13. Virginia Tower Norwood has sometimes been called "the woman who brought us the world" for her brilliant scientific instrument designs, which included a method for radar to track weather balloons (her solution: a small construction of disks spinning in the wind that created a characteristic pulsing signal in reflected radar data), other types of tracking antennae, the transmission system for the lunar Surveyor missions, and the famous multispectral camera for Earth observations. See, for example, this article on her work: Alice Dragoon, "The Woman Who Brought Us the World," *MIT Technology Review*, June 29, 2021, www.technologyreview.com/2021/06/29/1025732/the-woman-who-brought-us-the-world.

14. Landsat's early story is told in *Landsat's Enduring Legacy: Pioneering Global Land Observations from Space* (Landsat Legacy Project Team, 2022), https://my.asprs.org/landsat.

15. For information about the global monetary savings from Landsat, see "Orbiting Earth More Than 400 Miles Away in Space, Far from Human View," USGS, January 14, 2015, www.usgs.gov/news/featured-story/landsat-seen-stunning-return-public-investment.

16. The Google Earth Engine is worth exploring, even if you're not technically inclined, just to get a sense of the scope and value of all of this Earth information: https://earthengine.google.com.

17. The LAGEOS satellites are seldom talked about today, but not only are they two of the most beautiful satellites ever launched (spheres bejeweled with their fused silica reflectors), they are also still used for laser ranging, with the data supporting long-term study of Earth's rotation and shape. See, for example, "LAGEOS: LAser GEOdynamic Satellite," NASA, https://lageos.gsfc.nasa .gov. Sagan's message (etched on stainless-steel sheets inside the satellite) includes images depicting Earth's plate tectonics and the drift of continents over hundreds of millions of years.

18. You can find more information about the US GPS system here: "GPS: The Global Positioning System," www.gps.gov. In addition to this system there are, at present, three other satellite-navigation systems operating: Russia's GLONASS, the European Union's Galileo system, and China's BeiDou system.

19. The mission of OCEARCH represents a beautiful and important use of GPS tracking. You can go online and see what hundreds of large aquatic animals are doing, effectively in real time: www.ocearch.org/tracker. It's astonishing.

20. See, for example, A. Tatem, S. Goetz, and S. Hay, "Fifty Years of Earth-Observation Satellites," *American Scientist* 96 (2008): 390, www .americanscientist.org/article/fifty-years-of-earth-observation-satellites.

21. See "Actual Number of Objects Launched into Space," Our World in Data, https://ourworldindata.org/grapher/yearly-number-of-objects-launched -into-outer-space.

22. The risks posed by this kind of runaway scenario are still being debated (as is the likelihood of the scenario actually occurring). The original work is D. Kessler and B. Cour-Palais, "Collision Frequency of Artificial Satellites: The Creation of a Debris Belt," *Journal of Geophysical Research: Space Physics* (1978), https://doi.org/10.1029/JA083iA06p02637.

23. D. Murphy et al., "Metals from Spacecraft Reentry in Stratospheric Aerosol Particles," *Proceedings of the National Academy of Sciences* 120 (2023), https://doi.org/10.1073/pnas.2313374120. See also J. Ferreira et al., "Potential Ozone Depletion from Satellite Demise During Atmospheric Reentry in the Era of Mega-constellations," *Geophysical Research Letters* 51 (2024), https://doi .org/10.1029/2024GL109280.

24. See the Finnish company site, www.wisaplywood.com/wisawoodsat, and a description of the Japanese effort, https://en.wikipedia.org/wiki/LignoSat.

25. Jonathan Weiner, *The Beak of the Finch: A Story of Evolution in Our Time* (New York: Knopf, 1994), https://doi.org/10.5962/p.356855.

26. See Charles Darwin, *The Variation of Animals and Planets Under Domestication* (John Murray, 1875), https://darwin-online.org.uk/EditorialIntroductions/Freeman_VariationunderDomestication.html; and "Pigeons and Variation," https://darwin200.christs.cam.ac.uk/pigeons.

CHAPTER 5: TO DISTANT SHORES

1. A wonderfully detailed description of Luna 3 and its technology is available at mentallandscape.com, a site curated by Don P. Mitchell: http://mentallandscape.com/L_Luna3.htm.

2. The use of captured radiation-robust US-made film (from spy balloons that were a lot less radar invisible than hoped) on Luna 3 and the first Soviet images of the lunar farside are nicely described here: Alan Bellows, "Faxes from the Far Side," *Damn Interesting*, October 2015, www.damninteresting.com/faxes-from-the-far-side.

3. The final, complete NASA report on the Mariner 2 (Venus) mission and the scientific results is available here: "Mariner Venus 1962 Final Project Report," www.scribd.com/doc/45903256/Mariner-Venus-1962-Final-Project-Report.

4. I wrote about the Soviets "misplacing" Venus in this 2020 article: Caleb Scharf, "We Never Know Exactly Where We're Going in Outer Space," *Nautilus*, November 4, 2020, https://nautil.us/we-never-know-exactly-where-were-going-in-outer-space-238021.

5. Gemini 4 is mostly remembered for being the mission with the first successful spacewalk (EVA) by a US astronaut (Ed White), but the attempted rendezvous maneuver was perhaps equally important to the US space program at the time. You can follow along in the official transcript of all communications during the mission: https://historycollection.jsc.nasa.gov/JSCHistoryPortal/history/mission_trans/gemini4.htm.

6. Lagrange points represent an extremely important class of orbital phenomena, but they're not easy to gain an intuitive understanding of. In my years spent teaching these things at an advanced undergraduate level, I don't think I ever managed to do more than come up with easy-to-grasp explanations for the L1 and L2 points. (L1 is the balance of forces between the large masses, plus a "force" caused by orbital motion, and L2 can be thought of as a point that experiences the combined gravitational pull of both large masses, so its orbital period is faster than otherwise expected.) It's just hard to wrap your mind around a system in constant rotational motion with competing forces.

7. Halo orbits (see https://en.wikipedia.org/wiki/Halo_orbit) are one class of counterintuitive orbits that exist around Lagrange points. Predicting and mapping these require careful mathematical modeling.

8. Moons of moons are a fun subject, but the Roche limit is unforgiving. However, the limit is very dependent on the material properties—a fact later pointed out in more detail in George Darwin, "On the Figure and Stability of a Liquid Satellite," *Philosophical Transactions of the Royal Society of London* (Series A, Containing Papers of a Mathematical or Physical Character) 206 (1906): 161–248, http://doi.org/10.1098/rsta.1906.0018.

9. There's a great interview with Gary Flandro by David Swift in Swift's *Voyager Tales: Personal Views of the Grand Tour* (American Institute of Aeronautics and Astronautics, 1997), https://doi.org/10.2514/4.868931.

10. The grand tour and the incredible Pioneer and Voyager missions have inspired many books and articles; an excellent firsthand account is Jim Bell, *The Interstellar Age: Inside the Forty-Year Voyager Mission* (New York: Dutton, 2015).

11. See Carl Sagan, *Pale Blue Dot: A Vision of the Human Future in Space* (New York: Random House, 1994). Sagan's words included these famous lines: "Look again at that dot. That's here. That's home. . . . On it everyone you love, everyone you know, everyone you ever heard of, every human being who ever was, lived out their lives. The aggregate of our joy and suffering . . . every saint and sinner in the history of our species lived there—on a mote of dust suspended in a sunbeam."

12. The passage of Voyager 1 into the beginnings of the ocean of interstellar space (containing the so-called interstellar medium) was deduced by some major changes in what the spacecraft instruments registered. See, for example, W. Webber and F. McDonald, "Recent Voyager 1 Data Indicate That on 25 August 2012 at a Distance of 121.7 AU from the Sun, Sudden and Unprecedented Intensity Changes Were Observed in Anomalous and Galactic Cosmic Rays," *Geophysical Research Letters* 40 (2013), https://doi.org/10.1002/grl.50383. See also Ken Croswell, "Voyager Still Breaking Barriers Decades After Launch," *PNAS*, April 21, 2021, www.pnas.org/doi/10.1073/pnas.2106371118.

13. The JUICE mission to Jupiter really does have a literally circuitous route; you can see the present details of this at the European Space Agency: https://sci.esa.int/web/juice/-/58815-juices-journey-to-jupiter.

14. The specifications for cubesats were drawn up in 1999 as a way to make satellite experimentation and training easier. See "CubeSat," *Wikipedia*, https://en.wikipedia.org/wiki/CubeSat.

CHAPTER 6: THE RED SIREN

1. JPL's history in space exploration is pretty much second to none. An astonishing list of spacecraft missions, scientific instruments, and ingenious mathematical-physics methods have come from this laboratory. Many books have been written about that history, and JPL itself (seldom shying away from enthusing about its own reputation) hosts an excellent online resource: "History," www.jpl.nasa.gov/who-we-are/history.

2. An account of the making of the Mariner 4 pastel image is at "First TV Image of Mars," www.directedplay.com/first-tv-image-of-mars. This NASA movie of engineers making the map is also worth a watch: "Coloring the Image Data, in Color," NASA, https://science.nasa.gov/resource/coloring-the-image-data-in-color.

3. The Lyndon B. Johnson statement about Mariner 4's implications for life in the universe appears in this video: *JPL and the Space Age: The Changing Face of Mars*, NASA, 2023, https://plus.nasa.gov/video/jpl-and-the-space-age-the-changing-face-of-mars.

4. The Planetary Society has a useful list of every mission ever sent to Mars (both successful and failed): "Every Mission to Mars Ever," Planetary Society, www.planetary.org/space-missions/every-mars-mission.

5. The PrOP-M rover is often forgotten, but it was a bona fide attempt to place a near-foolproof mobile device on the surface of a world that was farther away than had ever been attempted before (other than the Moon). The acronym works only in Russian but stands for "Device Evaluation Terrain—Mars." The rover carried a dynamic penetrometer and a radiation densitometer, and it was intended to be put on the surface by a manipulator arm and then move in front of the lander's TV cameras. Its impressions in the soil would also provide information about consistency.

6. Mariner 9 returned an impressive 6.75 gigabytes of scientific data (a huge amount for the early 1970s), including 7,329 images that covered the entire planetary surface. After using up its supply of attitude-control gas, the spacecraft was turned off on October 27, 1972. It may still be orbiting Mars: it was estimated to have about fifty years before orbital decay would cause it to enter the atmosphere.

7. The Martian dichotomy is not fully understood, but many ideas exist. For one explanation (and a good summary of other ideas), see K. Cheng et al., "Mars's Crustal and Volcanic Structure Explained by Southern Giant Impact and Resulting Mantle Depletion," *Geophysical Research Letters* (2024), https://doi.org/10.1029/2023GL105910.

8. Extensive (mostly technical) information about the Viking 1 and 2 missions can be found at the NASA Space Science Data Coordinated Archive (NSS-DCA): https://nssdc.gsfc.nasa.gov/planetary/viking.html.

9. The strange results of the Viking biological experiments made a lot more sense after 2008, when the existence of highly reactive oxidizing perchlorate compounds were confirmed in the Martian regolith (soil), although a few scientists have continued to suggest that life was indeed responsible for the Viking results.

10. A comprehensive but technical review of Mars's past climate states is found in R. Wadsworth, "The Climate of Early Mars," *Annual Review of Earth and Planetary Sciences* 44 (2016), https://doi.org/10.1146/annurev-earth-060115-012355.

11. The Mars surface radiation environment is complex, and until fairly recently the data were sparse. A number of studies using in situ measurements are continuing. See, for example, D. Matthia et al., "The Martian Surface Radiation Environment—A Comparison of Models and MSL/RAD Measurements," *Journal of Space Weather and Space Climate* 6 (2016), https://doi.org/10.1051/swsc/2016008.

12. The current impact rate of small pieces of asteroid material on Mars has been evaluated from seismic data obtained with the InSight lander, which suggest that the rate may be two to ten times higher than previously estimated. See I. Daubar et al., "Seismically Detected Cratering on Mars: Enhanced Recent Impact Flux?," *Science Advances* 10 (2024), https://doi.org/10.1126/sciadv.adk7615. But more concerning are estimates of impacts by much larger objects, based on data on asteroid orbits and models, suggesting that Mars may have a two to three times higher probability of being hit: Y. Zhou et al., "MARTIANS (MARs2020, TIANwen and So On) Would See More Potentially Hazardous Asteroids Than Earthlings," *Monthly Notices of the Royal Astronomical Society: Letters* 532 (2024), https://doi.org/10.1093/mnrasl/slae040.

13. Terraforming is a subject rife with wild speculation and wild application of scientific ideas that are only approximations of how the world really works (leaving inconvenient details out). Nonetheless, it has a long history of discussion (and some merits in terms of the questions it poses), even by sober researchers like Carl Sagan. See, for example, his article "Planetary Engineering on Mars," *Icarus* 20 (1973), https://doi.org/10.1016/0019-1035(73)90026-2.

14. Elon Musk's statements about what human settlement on Mars would look like are varied and are scattered across speeches, news articles, and social media. On January 17, 2020, on Twitter (before it was X), he mentioned that Mars "needs to be such that anyone can go if they want, with loans available

for those who don't have money," leaving open the question of how such a loan would be paid back.

15. Carson Teuscher, "The Cold, Cold War: Rear Admiral Richard Byrd, Antarctic Expeditions, and the Evolution of America's Strategic Interest in the Polar Regions," Arctic Institute, November 2, 2021, www.thearcticinstitute.org /cold-cold-war-rear-admiral-richard-byrd-antarctic-expeditions-evolution -americas-strategic-interest-polar-regions.

16. The MOXIE experiment on Mars: J. Hoffman et al., "Mars Oxygen ISRU Experiment (MOXIE)—Preparing for Human Mars Exploration," *Science Advances* 8 (2022), https://doi.org/10.1126/sciadv.abp8636.

17. I wrote about the problem of human garbage on Mars in "The Garbage Dumps of Mars," *Slate*, July 5, 2022, https://slate.com/technology/2022/07 /mars-colonization-garbage-recycling.html.

18. See, for example, A. Campelo dos Santos et al., "Genomic Evidence for Ancient Human Migration Routes Along South America's Atlantic Coast," *Proceedings of the Royal Society B* 289 (2022), http://doi.org/10.1098/rspb.2022.1078.

19. I'm cautious about too many sweeping assertions about how racist and prejudiced many "great people" were in the past, but it's impossible to read some of Darwin's statements without seeing precisely that. Furthermore, behind that story is another story of who the Fuegians really were and what they had to deal with (both in terms of local environment and the appalling consequences of interactions with people from outside). This 2021 opinion essay presents an interesting analysis: Josie Glausiusz, "Savages and Cannibals: Revisiting Charles Darwin's Voyage of the Beagle," *Emerge*, www.whatisemerging.com/opinions/savages-and-cannibals.

20. Ray Bradbury's *The Martian Chronicles* (New York: Doubleday, 1950) was also published as *The Silver Locusts* in the United Kingdom, the title reflecting the notion of a locust-like devastation wrought on Mars by the occupants of the silver rocket ships coming from Earth.

CHAPTER 7: THE MONUMENTS AT THE EDGE OF THE BELIEVABLE

1. The total mass of the present technosphere has been estimated at about 30 trillion tonnes (metric), which is still a tiny percentage of the mass of the Earth. See J. Zalasiewicz et al., "Scale and Diversity of the Physical Technosphere: A Geological Perspective," *Anthropocene Review* 4, no. 1 (2017): 9–22, https://doi.org/10.1177/2053019616677743.

2. I wrote about Jupiter's gravitational pull on the Earth in a 2011 essay: Caleb A. Scharf, "Jovian Attraction," *Scientific American*, November 21, 2011, www.scientificamerican.com/blog/life-unbounded/jovian-attraction.

3. Pioneer 10's encounter with Jupiter was a major event in space exploration as the first "up-close" engagement with the worlds of the outer solar system. See, for example, "Pioneer 10," NASA, https://science.nasa.gov/mission/pioneer-10. This encounter resulted in more than five hundred images of the Jovian system. See R. Fimmel et al., "Pioneer: First to Jupiter, Saturn and Beyond," NASA special publication 446, 1980, https://atmos.nmsu.edu/data_and_services/atmospheres_data/SATURN/logs/nasa-sp-446-Pioneer-First-to-Jupiter-Saturn-and-Beyond.pdf.

4. Morabito has written a very thorough historical perspective on the discovery of Io's active volcanism and the nature of scientific discovery: Linda Morabito, "Discovery of Volcanic Activity on Io: A Historical Review," *arXiv*, 2012, https://arxiv.org/abs/1211.2554.

5. Europa's water-ice surface (and Ganymede's) was first documented by Gerard Kuiper in 1957 by using infrared spectra taken from the Earth.

6. Evidence for the salinity of Europa's interior ocean comes from a variety of research, some more theoretical, some using data obtained by Earth-based telescopes and space telescopes. See, for example, S. Trumbo, M. Brown, and K. Hand, "Sodium Chloride on the Surface of Europa," *Science Advances* 5 (2019), https://doi.org/10.1126/sciadv.aaw7123.

7. Amalthea is thought to likely be a "rubble pile" of quite loosely bound material with a higher ice content than expected. Its porous, icy composition suggests a complex history and perhaps that this moon is a captured asteroid.

8. The outer, irregular moon Himalia has only ever been photographed as a few pixels in an image, so we have limited information about its appearance, although it appears to be gray and to have a spin that gives it a day length of some seven hours. See, for example, F. Pilcher et al., "Photometric Lightcurve and Rotation Period of Himalia (Jupiter VI)," *Icarus* 219 (2012), https://doi.org/10.1016/j.icarus.2012.03.021.

9. The Galileo mission was a huge accomplishment, and the spacecraft itself is a bulky beast. If you visit JPL, you can view a full-scale model that really impresses with its 9-meter length and 4.6-meter diameter (with the high-gain antenna fully unfurled). See, for example, Michael Meltzer, "Mission to Jupiter: A History of the Galileo Project," NASA, 2013, www.nasa.gov/wp-content/uploads/2023/04/sp-4231.pdf.

10. An excellent technical review of the state-of-the-art understanding of Jupiter's interior, driven by Juno measurements, is D. Stevenson, "Jupiter's Interior as Revealed by Juno," *Annual Review of Earth and Planetary Sciences* 48 (2020), https://doi.org/10.1146/annurev-earth-081619-052855.

11. At the time of this writing, both JUICE and Europa Clipper are working well and are on their way to Jupiter after successful launches.

12. As with the Galileo mission, there are so many scientific articles and historical pieces about the Cassini-Huygens mission that it's hard to know what to refer readers to—there is a vast trove of research and discovery—but one starting point would be this book: Michael Meltzer, *The Cassini-Huygens Visit to Saturn: An Historic Mission to the Ringed Planet* (Cham: Springer, 2015).

13. For a scientific review of many aspects of the icy moon Enceladus, see J. Spencer and F. Nimmo, "Enceladus: An Active Ice World in the Saturn System," *Annual Review of Earth and Planetary Sciences* 41 (2013), https://doi .org/10.1146/annurev-earth-050212-124025.

14. As important as it is to understand Titan's surface environment, this environment is intimately connected to Titan's history and interior structure, which is also a subject of intense study. See, for example, C. Sotin et al., "Titan's Interior Structure and Dynamics After the Cassini-Huygens Mission," *Annual Review of Earth and Planetary Sciences* 49 (2021), https://doi.org/10.1146 /annurev-earth-072920-052847.

15. Low-temperature metabolic processes on Titan have been examined in some detail. See, for example, C. McKay and H. Smith, "Possibilities for Methanogenic Life in Liquid Methane on the Surface of Titan," *Icarus* 178 (2005), https://doi.org/10.1016/j.icarus.2005.05.018.

16. Dragonfly is a very ambitious mission, but it is also an entirely logical advancement of known technology, albeit in very alien conditions. For some scientific depth, see J. Barnes et al., "Science Goals and Objectives for the Dragonfly Titan Rotorcraft Relocatable Lander," *Planetary Science Journal* 2 (2021), https://iopscience.iop.org/article/10.3847/PSJ/abfdcf.

17. The tension between the natural and the unnatural is an enduring philosophical and ethical debate, and to a significant extent a scientific debate as well. If human actions can be categorized as "unnatural," isn't that separating us from everything else in the universe—a form of reverse Copernicanism or reversal of the post-Einsteinian cosmological principle? Sociobiology is an interesting subject matter, though. See E. O. Wilson, *Sociobiology: The New Synthesis* (Cambridge, MA: Belknap Press, 1975, and the later editions).

CHAPTER 8: THE INNERS AND THE OUTERS

1. The nebular hypothesis is also often referred to as the Kant-Laplace nebular hypothesis for star and planet formation. The idea was first published by Kant in 1755 (in his *Universal Natural History and Theory of the Heavens*) and

then independently considered and furthered by Laplace in 1796 (in *The System of the World*). Neither version got everything right, but the overall concept of gravitational collapse and condensation of matter to eventually form stars and planets was broadly correct.

2. The grand tack model for Jupiter's early history is an intriguing idea. See, for example, S. Raymond and A. Morbidelli, "The Grand Tack Model: A Critical Review," *Proceedings of the International Astronomical Union* 9 (2014): 194–203, https://doi.org/10.1017/S1743921314008254.

3. *Planet Bura* is pretty marvelous fun. It has it all, from rockets and zero-g to dinosaurs, robots, and human antics. See www.youtube.com/watch?v=-VTF23EFvEM.

4. One of the very best resources for learning about the details of the Soviet Venus missions (and other early space exploration) is in the website curated by Don P. Mitchell: http://mentallandscape.com/V_Venus.htm. Mitchell has also done excellent work on reconstructing and cleaning up some of the imaging data on Venus.

5. The Vega balloon (aerostat) experiments were a real triumph. See, for example, J. Blamont, "The VEGA Venus Balloon Experiment," *Advances in Space Research* 7 (1987), https://doi.org/10.1016/0273-1177(87)90233-X.

6. The history and evolution of the Venusian climate is a topic of ongoing study. It is possible that Venus was (in the early solar system) much more "Earth-like" in terms of temperate conditions. See, for example, C. Gilmann et al., "The Long-Term Evolution of the Atmosphere of Venus: Processes and Feedback Mechanisms," *Space Science Reviews* 218 (2022), https://doi.org/10.1007/s11214-022-00924-0. See also M. Way and A. Del Genio, "Venusian Habitable Climate Scenarios: Modeling Venus Through Time and Applications to Slowly Rotating Venus-Like Exoplanets," *Journal of Geophysical Research: Planets* 125 (2020), https://doi.org/10.1029/2019JE006276.

7. The best global radar maps of Venus came from the Magellan mission, which arrived at Venus in 1990 and operated until 1994. See "*Magellan* (Spacecraft)," *Wikipedia*, https://en.wikipedia.org/wiki/Magellan_(spacecraft).

8. The idea that organisms could exist in the upper parts of Venus's atmosphere (where pressures and temperatures are similar to those on the Earth's surface and solar energy is abundant) has been long debated. In recent years, claims of the detection of the compound phosphine in the atmosphere have led to suggestions that phosphine-producing microorganisms might be living in a cloud ecosystem. See, for example, S. Seager et al., "The Venusian Lower Atmosphere Haze as a Depot for Desiccated Microbial Life: A Proposed Life Cycle

for Persistence of the Venusian Aerial Biosphere," *Astrobiology Journal* 21, no. 10 (2020), https://doi.org/10.1089/ast.2020.2244). However, measurements and proposals remain highly controversial, with a dearth of good data to support or refute such possibilities.

9. Mercury is a deceptively challenging world to explore. For an overview of the challenges and the MESSENGER mission as a case study, see D. L. Domingue and C. T. Russell, eds., *The MESSENGER Mission to Mercury* (New York: Springer, 2007). See also "The MESSENGER Mission: Science and Implementation Overview," in *Mercury: The View After MESSENGER*, ed. S. C. Solomon, L. R. Nittler, and B. J. Anderson (Cambridge: Cambridge University Press, 2018), 1–29.

10. The fact that electromagnetic radiation exerts a pressure force was first properly quantified by the physicist James Clerk Maxwell in the 1860s with his theory of electromagnetic fields, demonstrating that light carries momentum and therefore (as with conventional rocket propulsion) can accelerate objects. With everyday light intensities on Earth the effect is tiny, but around 1900 it was experimentally verified.

11. For the time being, our insights about the planets Uranus and Neptune mostly come from studying their upper atmospheres from afar, but there is a lot that can be learned that way. See, for example, this review: H. Melin, "The Upper Atmospheres of Uranus and Neptune," *Philosophical Transactions of the Royal Society A* 378 (2020), http://doi.org/10.1098/rsta.2019.0478.

12. Triton is extremely interesting. When Voyager 2 encountered the Neptunian system in 1989, it was able to capture a handful of images of Triton that showed, in very crude time lapse, what appeared to be the shadowing clouds of cryo-volcanic "geysers" of dust and what was likely nitrogen and organic compounds erupting from the moon's surface to altitudes of several kilometers. See L. A. Soderblom et al., "Triton's Geyser-Like Plumes: Discovery and Basic Characterization," *Science* 250 (1990), http://doi.org/10.1126/science.250.4979.410.

13. The Kuiper belt is named after the astronomer Gerard Kuiper (1905–1973), who pioneered many aspects of modern planetary science and developed some of the early hypotheses about what lies beyond the orbits of Uranus and Neptune.

14. The story of the New Horizons mission and the encounter with Pluto is told in Alan Stern and David Grinspoon, *Chasing New Horizons: Inside the Epic First Mission to Pluto* (New York: Picador, 2018).

15. How Pluto got its neat orbit, which preserves it against collision or disruption by Neptune, is an interesting question that seems to have a

solution in some sophisticated computer modeling of orbital dynamics. See, for example, Renu Malhotra, "The Origin of Pluto's Orbit: Implications for the Solar System Beyond Neptune," *arXiv*, 1995, https://arxiv.org/abs/astro-ph /9504036.

16. The term "Kuiper belt object" (KBO) is often used interchangeably with the term "trans-Neptunian object" (TNO), although, strictly speaking, TNO is the overall "class" of all objects orbiting beyond Neptune (including minor planets like Pluto and Eris), whereas KBOs are a subset of those up to a distance of about fifty-five times farther from the Sun than the Earth is. Beyond that distance, objects are often classified as scattered disk objects (SDOs). I know that it's confusing: astronomers and planetary scientists don't always pick the easiest naming conventions. I've used "Kuiper belt object" a little loosely in order to minimize the number of terms.

17. A good short article about the Deep Space Network is available here: "The Deep Space Network: NASA's Link to the Solar System," Lunar and Planetary Institute, www.lpi.usra.edu/publications/newsletters/lpib/new/the-deep-space -network-nasas-link-to-the-solar-system.

CHAPTER 9: THE FURNACE

1. Carrington's observations, the associated events that took place around the world, and the implications for future solar storms have been written about in many places. See, for example, S. Odenwald and J. Green, "Bracing for a Solar Storm," *Scientific American*, August 2008. More-technical details are reported in E. Cliver and W. Dietrich, "The 1859 Space Weather Event Revisited: Limits of Extreme Activity," *Journal of Space Weather and Space Climate* 3 (2013), www .swsc-journal.org/articles/swsc/pdf/2013/01/swsc130015.pdf.

2. Carrington's own report for the Royal Astronomical Society in 1859 can be found at https://babel.hathitrust.org/cgi/pt?id=njp.32101081655332&view =1up&seq=358.

3. M. A. Shea et al., "Solar Proton Events for 450 Years: The Carrington Event in Perspective," *Advances in Space Research* 38 (2006), https://doi.org /10.1016/j.asr.2005.02.100.

4. The 775 CE event is also known as the 774–775 carbon-14 spike. The enhancement of the carbon-14 isotope seen in tree rings from this timespan across the globe indicates a bump about twenty times more than the usual year-to-year variation. Although there might be other possible explanations (such as more distant cosmic events), the general scientific consensus is that Earth experienced a very strong solar-particle event that produced excess carbon-14.

5. Rockoons—aerial rocket launches that carry the rocket above the lower atmosphere using balloons—worked well for the small sounding rockets that helped researchers scope out the outer reaches of Earth's atmosphere in the 1950s. See, for example, Colleen C. Anderson, "Rockoons: Rocket and Balloon Experiments," National Air and Space Museum, June 1, 2024, https://airandspace.si.edu/stories/editorial/rockoons-rocket-and-balloon-experiments.

6. D. Knipp et al., "On the Little-Known Consequences of the 4 August 1972 Ultra-fast Coronal Mass Ejecta: Facts, Commentary, and Call to Action," *Space Weather* 16 (2018), https://doi.org/10.1029/2018SW002024.

7. "Helios 1," NASA, https://science.nasa.gov/mission/helios-1.

8. The Solar and Heliophysics Observatory (SOHO) has been extraordinarily successful, generating a truly vast trove of data: https://soho.nascom.nasa.gov.

9. There is even a spin-off citizen-science project devoted to this task: https://sungrazer.nrl.navy.mil.

10. The autonomy of the Parker Solar Probe has to be particularly robust or else risk loss of the entire mission. For a technical discussion, see R. Smith et al., "Integration and Test Challenges of Parker Solar Probe," IEEE Aerospace Conference, 2020, https://doi.org/10.1109/AERO47225.2020.9172278.

11. Auroras have been seen in the atmospheres of all the outer giant planets.

12. The eleven-year solar cycle is a complicated phenomenon consisting of many features of the Sun, from sunspots to radio emissions and overall magnetic behavior. For a very complete technical review, see D. Hathaway, "The Solar Cycle," *Living Reviews in Solar Physics* 7, no. 1 (2010), https://doi.org/10.12942/lrsp-2010-1.

13. Radiation for humans in space has many ramifications for shielding and for timing exploration. For instance, in order to minimize radiation risks from the highest-energy cosmic radiation, the best time to launch to Mars may be during the solar *maximum*, and the mission should last no longer than four years: M. Dobynde et al., "Beating 1 Sievert: Optimal Radiation Shielding of Astronauts on a Mission to Mars," *Space Weather* 19 (2021), https://doi.org/10.1029/2021SW002749.

14. Many studies exist on materials that are best suited to shield against particle radiation. See, for example, S. Thibeault et al., "Radiation Shielding Materials Containing Hydrogen, Boron, and Nitrogen: Systematic Computational and Experimental Study—Phase I," NIAC Report, 2012, www.nasa.gov/wp-content/uploads/2017/07/niac_2011_phasei_thibeault_radiationshieldingmaterials_tagged.pdf.

15. The seminal paper on the discovery of radiation-induced errors in integrated circuits appeared in 1978: T. May and M. Woods, "A New Physical Mechanism for Soft Errors in Dynamic Memories," 16th International Reliability Physics Symposium, San Diego, CA, 1978, 33–40, https://doi.org/10.1109/IRPS.1978.362815.

CHAPTER 10: THE COUNTRY OF A BILLION SHIRES

1. Darwin wrote about what he was finding in the Galapagos with expressions of wonder, so I'm taking some interpretative liberty in also seeing the pain in his mental wrangling to try to explain the nature of species across these islands. He definitely expresses many different possible explanations without seeming to be able to pin anything down. As he writes, "Reviewing the facts here given, one is astonished at the amount of creative force, if such an expression may be used, displayed on these small, barren, and rocky islands; and still more so, at its diverse yet analogous action on points so near each other."

2. On a single, active planet like the Earth, even if life had never existed, the surface contents would have been overturned and changed again and again as chemistry and geophysics stirred and modified things. Adding life seems to throw fuel onto those processes, with phenomena building, mixing, and erasing other phenomena again and again—both the blessing and the curse of a finite planetary surface.

3. The dwarf planet Ceres was studied by the Dawn mission from orbit in 2015–2018, and the data it obtained really rewrote our conception of this world. It is far more complicated and interesting than we had perhaps expected, particularly because much of its history seems to have involved the effects of water. T. McCord et al., "Ceres, a Wet Planet: The View After Dawn," *Geochemistry* 82 (2021), https://doi.org/10.1016/j.chemer.2021.125745.

4. It can seem surprising that the first real encounter between a spacecraft and an asteroid was the Galileo-Gaspra encounter in 1991, but up until this point most nations were fixated on planetary quests. In many ways, asteroid and comet exploration is a twenty-first-century endeavor.

5. The European Space Agency's Rosetta mission produced some of the most spectacular images and data on a cometary body that we have seen to date. You can browse the one hundred thousand images here: https://imagearchives.esac.esa.int.

6. The Sakigake and Suisei spacecraft encounters with Halley's Comet are described here: K. Hirao and T. Itoh, "The Sakigake/Suisei Encounter with Comet p/ Halley," *Astronomy and Astrophysics* 187 (1987).

7. You can read about JAXA's audacious Hayabusa mission and some of its discoveries at "Hayabusa," JAXA, www.isas.jaxa.jp/en/missions/spacecraft/past/hayabusa.html; and "Topics: Archive," JAXA, www.isas.jaxa.jp/e/snews/2006/0602.shtml.

8. The acronym OSIRIS-REx stands for "Origins, Spectral Interpretation, Resource Identification, Security, Regolith Explorer," and the mission website contains a wealth of information: www.asteroidmission.org.

9. The Deep Impact autonomous targeting and control system is described here: M. Nikos et al., "Autonomous Navigation for the Deep Impact Mission Encounter with Comet Tempel 1," *Space Science Reviews* 117, nos. 1–2 (2005): 95–121, https://doi.org/10.1007/s11214-005-3394-4.

10. The full analysis of DART's effects on Dimorphos is in D. Richardson, "The Dynamical State of the Didymos System Before and After the DART Impact," *Planetary Science Journal* 5 (2024), https://iopscience.iop.org/article/10.3847/PSJ/ad62f5.

11. The possibility of ejected Dimorphos boulders eventually hitting Mars is described here: M. Fenucci and A. Carbognani, "Long-Term Orbital Evolution of Dimorphos Boulders and Implications on the Origin of Meteorites," *Monthly Notices of the Royal Astronomical Society* 528 (2024), https://doi.org/10.1093/mnras/stae464.

12. The mass of the technosphere is essentially on a par with the total dry biomass of life on Earth. See, for example, E. Elhacham et al., "Global Human-Made Mass Exceeds All Living Biomass," *Nature* 588 (2020), https://doi.org/10.1038/s41586-020-3010-5. Note also—as a reference point—that the total mass of buildings in New York City is estimated to be about 8×10^{11} kg distributed over a 780-square-kilometer area.

13. J. Syvitski et al., "Extraordinary Human Energy Consumption and Resultant Geological Impacts Beginning Around 1950 CE Initiated the Proposed Anthropocene Epoch," *Communications Earth and Environment* 1, no. 32 (2020), https://doi.org/10.1038/s43247-020-00029-y. This amounts to 22 zettajoules since 1950 alone, plus about 15 zettajoules across the last 12,000 years, totaling 37 zettajoules (37×10^{21} joules). Bear in mind that in 1831, when Darwin set out on the *Beagle*, the total human population was about 1 billion.

14. The energy release of the Sun in one hour is about 10^{31} joules, and the amount of that hitting the top of Earth's atmosphere averages at 4.3×10^{20} joules per hour. About three to four days' worth of solar hitting the top of the atmosphere would be the same energy as humanity has used in the past 12,000 years.

15. Michael Mautner, "Life in the Cosmological Future: Resources, Biomass and Populations," *Journal of the British Interplanetary Society* 58, no. 5 (2005): 167–180.

16. Ion-thruster technology continues to evolve. See, for example, this major review: K. Holste et al., "Ion Thrusters for Electric Propulsion: Scientific Issues Developing a Niche Technology into a Game Changer," *Review of Scientific Instruments* 91 (2020), https://doi.org/10.1063/5.0010134.

17. An interesting piece on the economics of rocket launches and its evolution is James Pethokoukis, "Moore's Law Meet Musk's Law: The Underappreciated Story of SpaceX and the Stunning Decline in Launch Costs," *Faster, Please!*, March 26, 2024, https://fasterplease.substack.com/p/moores-law-meet-musks -law-the-underappreciated. It's important to also understand that companies like SpaceX, Rocket Lab, and other commercial one-stop-shop commercial-launch systems have a much more flexible "risk posture" than do national agencies like NASA or ESA because they're less directly spending taxpayer money. That helps with the development of things like reusable, self-landing rockets, even though the basic technology has been understood for quite a long time. This doesn't diminish the accomplishments, but it does provide context.

18. A good article on why cheapness of launch is pivotal for the cost of all parts (including making low-mass, highly customized spacecraft to deal with launch limits): R. Simberg, "Walmart, but for Space," *New Atlantis*, 2021, www .thenewatlantis.com/publications/walmart-but-for-space.

19. See, for example, "Space Economy Set to Triple to $1.8 Trillion by 2035, New Research Reveals," World Economic Forum, April 8, 2024, www .weforum.org/press/2024/04/space-economy-set-to-triple-to-1-8-trillion-by -2035-new-research-reveals.

20. Gerard O'Neill was an interesting and exceedingly imaginative scientist. See, for instance, his *The High Frontier: Human Colonies in Space* (New York: William Morrow, 1976). See also "O'Neill Cylinder," *Wikipedia*, https: //en.wikipedia.org/wiki/O%27Neill_cylinder.

21. For some of the history of space habitat ideas, see J. Logsdon and G. Butler, "Space Station and Space Platform Concepts: A Historical Review," in *Space Stations and Space Platforms—Concepts, Design, Infrastructure, and Uses*, ed. I. Bekey and D. Herman (American Institute of Aeronautics and Astronautics, 1985), https://arc.aiaa.org/doi/book/10.2514/4.865749.

22. T. Maindl et al., "Stability of a Rotating Asteroid Housing a Space Station," *arXiv*, 2018, https://arxiv.org/abs/1812.10436.

CHAPTER 11: THE DISPERSAL

1. Atmospheric methane concentrations are less often in the news than carbon dioxide concentrations, but both can be tracked here: "Methane," NASA, https://climate.nasa.gov/vital-signs/methane/?intent=121.

2. This statement is widely attributed to Yuri Gagarin but is a little tricky to pin down. It may come from his book *Road to the Stars*. See Y. Gagarin, *Road to the Stars*, trans. G. Hanna and D. Myshe (Moscow: Foreign Languages Publishing House, 1962). I confess to not having managed to find a readable version of this. But even if the statement is an embellishment of something he said, it is very consistent with other comments or official statements that he made at the time.

3. Estimates of the total homo sapiens population a hundred thousand years ago are that it amounted to fewer than a million individuals and was probably more like a few hundred thousand at most, based on models using mutation rates in genetic data.

4. This hypothetical scenario is clearly very much based on current technological approaches to things like manufacturing (including 3D printing) and mobility with drones. A real future like this could use tools that we haven't come close to realizing yet.

5. The Hiten mission was quite audacious; it even included a smaller orbiter called Hagoromo (after the feather mantle of Hiten bodhisattvas) to be released directly into lunar orbit. However, the Hagoromo transmitters failed.

6. Ed Belbruno and James Miller, *Fly Me to the Moon: An Insider's Guide to the New Science of Space Travel* (Princeton, NJ: Princeton University Press, 2007). The technical report of Belbruno and Miller's work on the Hiten trajectory is available as E. Belbruno and J. Miller, "A Ballistic Lunar Capture Trajectory for the Japanese Spacecraft Hiten," Technical Report JPL-IOM 312/90.4-1731 -EAB, Jet Propulsion Laboratory, 1990.

7. The Interplanetary Transport Network idea comes from a number of mathematical works, including the weak stability boundary theory. See, for example, S. Ross, "The Interplanetary Transport Network," *American Scientist* (2006), https://doi.org/10.1511/2006.3.230. See also J. Marsden and S. Ross, "New Methods in Celestial Mechanics and Mission Design," *Bulletin of the American Mathematical Society* 43 (2006), https://doi.org/10.1090/S0273-0979-05-01085-2; and E. Belbruno, *Capture Dynamics and Chaotic Motions in Celestial Mechanics: With Applications to the Construction of Low Energy Transfers* (Princeton, NJ: Princeton University Press, 2004).

8. Towers and elevators to space are an alluring concept but also a fantastically tricky engineering problem—whether in the strengths of materials or

in all the associated problems to solve. Yuri Artsutanov's idea of a cable from geosynchronous anchors is perhaps the least difficult to implement. It was popularized in Arthur C. Clarke's 1979 novel *The Fountains of Paradise*. Many (many) papers and analyses have been produced over the years. See, for example, P. K. Aravind, "The Physics of the Space Elevator," *American Journal of Physics* 45, no. 2 (2007): 125; and J. Pearson, "The Orbital Tower: A Spacecraft Launcher Using the Earth's Rotational Energy," *Acta Astronautica* 2, nos. 9–10 (1975): 785–799.

9. Lunar space elevators aren't just for the Moon. See, for example, Z. Penoyre and E. Sandford, "The Spaceline: A Practical Space Elevator Alternative Achievable with Current Technology," *arXiv*, 2019, https://arxiv.org/abs/1908.09339.

10. J. Isaacs et al., "Satellite Elongation into a True 'Sky-Hook,'" *Science* 151 (1966), https://doi.org/10.1126/science.151.3711.682.

11. "Hypersonic Airplane Space Tether Orbital Launch (HASTOL) Architecture Study," Boeing, 2001, www.niac.usra.edu/files/studies/final_report/391Grant.pdf.

12. Y. Tsuda et al., "Achievement of IKAROS—Japanese Deep Space Solar Sail Demonstration Mission," *Acta Astronautica* 82 (2013), https://doi.org/10.1016/j.actaastro.2012.03.032.

13. For an example of laser-propulsion challenges, see Z. Manchester and A. Loeb, "Stability of a Light Sail Riding on a Laser Beam," *Astrophysical Journal Letters* 837 (2017), https://doi.org/10.3847/2041-8213/aa619b.

14. The idea of 3D printing or additive manufacturing seems to have first appeared in Western science fiction in the 1940s with loosely described machines that "construct" things from raw ingredients. The first real techniques were pioneered in the 1970s and 1980s, with metals and UV-hardening polymers, and in the mid-2000s the designs and instructions for "homemade" 3D printers began to circulate. The largest 3D polymer printers today can make structures up to about thirty meters in length, and concrete-extrusion systems can produce entire buildings.

15. D. R. Colman et al., "The Deep, Hot Biosphere: Twenty-Five Years of Retrospection," *PNAS* 114, no. 27 (2017): 6895–6903, https://doi.org/10.1073/pnas.1701266114; and G. Borgonie et al., "New Ecosystems in the Deep Subsurface Follow the Flow of Water Driven by Geological Activity," *Scientific Reports* 9 (2019), https://doi.org/10.1038/s41598-019-39699-w.

INDEX

Nix, 128, 223
Noachian period, 151
Noachis Terra (Land of Noah), 151
Noether, Amalie Emmy, 33–34, 41, 53, 282
Noether's Theorem, 34, 55
noösphere, 262
Norwood, Virginia Tower, 92–93
Nozomi orbiter, 149

Oberth, Hermann, 53, 83
Oberth effect, 53–56, 53*fig*, 266
OCEARCH, 102
octocopter, 201–202
Olympus Mons, 146–147, 152
On the Origins of Species (Darwin), 15, 229
on-demand manufacture, 288
O'Neill, Gerard, 273–274
O'Neill cylinders, 273–274, 275
Opportunity rover, 150
orbital hierarchies, 127–128
orbital mechanics, 122–123
orbital momentum, 130–131
orbital rendezvous, 121–123
Orbiting Solar Observatory Program (OSO), 236
orbits
    around Moon, 67–68
    asteroids and, 258–259
    changes in, 55
    closed, 37, 120
    energy and, 37
    frozen, 67–68
    geostationary, 129
    Hohmann transfer and, 38, 39*fig*
    Hohmann transfer orbit, 123–124
    human modification of, 7

mean-motion resonance and, 182, 183
retrograde, 221
stable, 128
sun-synchronous, 129
sustainable, 21, 35–37, 36*fig*
trailing, 299
See also satellites
Ordo Templi Orientis, 52
orienting spacecraft, 115–116
OSIRIS-REx, 257, 258
overview effect, 108–109, 110–111
oxygen, Mars and, 163–164

"Pale Blue Dot," 134–135
Paradise Lost (Milton), 13
Parker Solar Probe, 237–238
Parker Spiral, 242
Parsons, Jack, 51–52
perigee, 55
Perseverance rover, 150, 163
perturbations, theory of, 32–33
Pettit, Don, 49
PFS-1 and PFS-2, 67–68
Phobos (moon), 155, 257
Phobos missions, 149
Phoenix lander, 150
photodetectors, 116
pigeons, evolution and, 110
Pioneer missions, 132–133, 135, 182, 188, 193, 197, 202, 239, 253
Pioneer Venus 1 and 2, 212
place, conceptions of, 35
Plan for Space Exploration (Tsiolkovsky), 40
Planet Bura (*Planet of Storms*), 209
Planet Labs, 95
planetary hierarchies, 128

on Jupiter's moons, 183–186, 187
Moon and, 73–74, 78–79
navigation and, 120–121
recycling of on ISS, 164
satellite information and, 94
on Saturn's moons, 196–197
Watson, James, 15
weak stability boundary theory, 288
wealth gap, 106, 161
weather satellites, 6, 91–92, 111
weightlessness, effects of, 82–83
Weiner, Jonathan, 110
Wells, H. G., 52
White, Ed, 121
Wilkins, Maurice, 15

Wisconsin Glaciation, 62
wood, satellites made from, 107

X-rays, 244

Yarkovsky effect, 218
YORP effect, 218
Yucatan Peninsula, asteroid collision
  and, 254

Zeno's paradox, 42
Zhurong rover, 150
Zond 5, 82
zones of easiest exploration, 226, 227,
  270–271

Credit: Nerissa Escanlar

Caleb Scharf received the 2022 Carl Sagan Medal while director of astrobiology at Columbia University and is currently the senior scientist for astrobiology at NASA's Ames Research Center. He has written several previous trade books and is a frequent contributor to *Scientific American* and *Nautilus* magazine. He divides his time between Silicon Valley and New York City.

RAISING READERS
Books Build Bright Futures

Thank you for reading this book and for being a reader of books in general. As an author, I am so grateful to share being part of a community of readers with you and I hope you will join me in passing our love of books on to the next generation of readers.

Did you know that reading for enjoyment is the single biggest predictor of a child's future happiness and success?

More than family circumstances, parents' educational background, or income, reading impacts a child's future academic performance, emotional well-being, communication skills, economic security, ambition, and happiness.

Studies show that kids reading for enjoyment in the US is in rapid decline:

- In 2012, 53% of 9-year-olds read almost every day. Just 10 years later, in 2022, the number had fallen to 39%.
- In 2012, 27% of 13-year-olds read for fun daily. By 2023, that number was just 14%.

Together, we can commit to **Raising Readers** and change this trend. How?

- Read to children in your life daily.
- Model reading as a fun activity.
- Reduce screen time.
- Start a family, school, or community book club.
- Visit bookstores and libraries regularly.
- Listen to audiobooks.
- Read the book before you see the movie.
- Encourage your child to read aloud to a pet or stuffed animal.
- Give books as gifts.
- Donate books to families and communities in need.

Books build bright futures, and **Raising Readers** is our shared responsibility.

For more information, visit **JoinRaisingReaders.com**

Sources: National Endowment for the Arts, National Assessment of Educational Progress, WorldBookDay.org, Nielsen BookData's 2023 "Understanding the Children's Book Consumer"